QUANTITATIVE METHODS IN GEOGRAPHICAL RESEARCH

QUANTITATIVE METHODS IN GEOGRAPHICAL RESEARCH

Dr. (Mrs.) Najma Khan

CONCEPT PUBLISHING COMPANY, NEW DELHI-110059

ISBN 81-7022-696-1 (HB)

ISBN 81-7022-697-X (PB)
First Published 1998
Reprinted 2002

Published and Printed by
Ashok Kumar Mittal
Concept Publishing Company
A/15-16, Commercial Block, Mohan Garden
New Delhi-110059 (India)

Phones : 5554042, 5504042
Fax 011-91-5598898

Preface

The monograph entitled *Quantitative Methods in Geographical Studies* is intended primarily for the students of Undergraduate and Postgraduate of Social Sciences and Humanities as they do not have much background in statistics. Hence, most of the statistical concepts and methods have been supplemented by easy examples and exercises. I have also tried to show the relevance of statistical techniques to geographic problems through the use of data made available from toposheets such as contour data for slope analysis and rivers for the intensity and density of rivers to measure the significance of these rivers for water resources. Similarly, the data for the statistical analysis have been taken from various original sources for population, agriculture, settlements and geomorphology.

Most of the ideas in the book have arisen while teaching a number of courses and their practicals. It was realised that since there is no book available to tackle such problems. For instance, if anyone working or studying the field of population geography he should get most of the statistical techniques and methods in a single volume. Similarly, the fields of agricultural geography, settlement geography and geomorphology was covered and all the possible statistical techniques have been covered in this book.

Many people and literature sources helped, directly or indirectly, with this book. All my colleagues at the Department in Egerton University, have provided a very pleasant atmosphere. In particular, Mrs. J.P. Musangi commented on various parts of an early draft of the manuscript, and Mrs. Rose Marry Hayanga, J. Otieno, Mwasi and I.K. Rop kindly made available sets of data for the courses being taught by them. Prof. Mrs. J.P. Musangi and Prof. Ali Mohammad provided many papers giving examples of statistical applications in physical geography. I am grateful to all these colleagues who patiently explained the background of some of the exercises. A number of colleagues of other departments also very kindly discussed a number of statistical issues with me at some length, although of course all responsibility for the particular interpretations presented and any error, are mine alone.

I am also grateful particularly to B.L. Agarwal and Aslam Mahmood who wrote *Basic Statistics* and *Statistical Methods in Geographical Studies* respectively. These two books inspired me significantly to complete the present work. The author is highly thankful to many of the authors engaged in this field.

DR. NAJMA KHAN
Senior Lecturer,
Egerton University,
Kenya.

Contents

Introduction

The study of geography, along with other social science subjects, has passed through several stages of changes. Recently it was realised that it is better to make the results more and more convincing and reliable because a simple generalised picture may not help in understanding the subject. During the stages of planning of various socio-economic aspects the data supplement with sound base of reliability helps in making the planning very sound. Hence, the quantification started both for the collection and analysis of data.

For achieving accuracy, precision and depth of the quality of any region the analysis of various indicators is very necessary such as land resource and land capability index, aridity index, per head income and consumption of food, workforce in secondary sector of economy, index of urbanization, literacy rate, index of income inequality, percent share of deprived persons in the total population, working women or participation in the total female population. The values of these indicators either singly or after composition help him to arrive at a qualitative assessment about the health of the region.

The geographers have been using quantitative methods for identification, explanation and decision-making. Their attempts in this direction generally take the forms such as : to collect and tabulate data methodically and systematically, to extract sample from a large and unmanageable universe in such a manner that the analysis with the sample become valid and representative for the universe, to identify, classify and extract the inherent characteristics of phenomena, to identify association and correlation between and combination of phenomena across space and time, to study distributions (frequency, measures of inequality, growth and development, spatio and temporal variations, concentration, clustering and dispersion), to explain and relate causes and effects, to project processes in time and predict and to stimulate and build spatial models.

It is evident from the existing and continuing trend that geography is moving from descriptive to analysis and producing more reliable and acceptable results through the help of quantitative tools. But one should also be aware with some of the demerits and serious consequences.

It is very difficult rather impossible to select the indicators to measure the development. For instance the same indicator may be playing a negative role in the development of developing countries and positive in the development of the developed countries. As it is well known that the Colin Clark hypothesis of the share of the tertiary sector as an indicator of economic development is just not tenable in most of the developing economies wherein it is an indicator of disguised unemployment rather than of economic development. The dominance of tertiary activities in the Indian Urbanization process is more an indicator of the dysfunctional character of urbanization, as the growth of centres of consumption rather than production.

Secondly, selecting the indicators of development depends largely on the concept of the problems and how far the concept is clear and well understood is a matter of doubt especially with the social scientists of the developing countries? If the observational part is poor and structure of the area is misunderstood the collection and generating the sources of data may lead to a faulty and misuse of statistics and quantification. Therefore, observational part must be given more importance rather than statistical part. In one of the very important study of correlation it was stated and was tried to correlate the higher concentration of scheduled caste with the higher concentration of agricultural labourers. But the result obtained was not positive because the region was under the predominance of the concentration of scheduled caste population. Similarly a doctoral thesis proved, on the basis of computer intensive exercise that urban centres in India and the United States of America have a common feature, i.e., their populations are larger than those of their rural counterparts.

It is the objective of the quantitative methods to validate the theory, not to generate the theory and causal relationship should be rooted in logic. For example if two examples are taken for analysis: higher the amount of rainfall higher the productivity of foodgrains as well higher the rate of irrigation higher will be the productivity of foodgrains. Former case may or may not be related at various regional level while the latter one is related in all the cases.

How far one is convinced that the following variables or indicators are negatively or positively related with the development such as : income per head, literacy rate, manufacturing and registered factory workers, number of electricity establishments, roads per square kilometre, irrigated area are positively related with development, on the other hand, scheduled castes and tribes, hired workers, agricultural labourers, proportion of labourers and cultivators and rural population are negatively related with development.

The inherent danger to use the quantitative techniques should not direct the social scientists, especially the geographers, to use this discipline. Therefore, in order to maintain the proper interaction with other disciplines the use of statistics is necessary. One must keep in mind that a sophisticated quantitative technique needs sophisticated use of the collection of data with clear concept of the problem.

The present work, therefore, starts with the analysis of the various sources of data and data collection with data gap and data pitfalls as well as how far the data is reliable for the existing problems located in our socio-economic problems? Secondly, it deals with the use of statistics in geomorphology because the use of statistics in this field is a recent and extremely important phenomena. The use of statistics or quantitative revolution in the fields of population geography, settlement geography and agricultural geography have been discussed at length because these are the core of geography. Lastly, some of the important techniques have been discussed separatcly as they can be used in many types of studies related to this field of geography and other social sciences.

1

Definition, Objectives and Fields of Statistics

For the last few centuries, statistics has remained a part of mathematics as the original work was done by mathematicians like Pascal (1623-1662), James Bernoulli (1654-1705), Laplace (1749-1827), Gauss (1777-1855), Lagrange, Bayes, Markoff, Euler etc. These mathematicians were mainly interested in the development of the theory of probability as applied to the theory of games and other chance phenomena. Till the early nineteenth century, statistics was mainly concerned with official statistics, needed for the collection of information on revenue, population and area of land under cultivation etc., of a state or kingdom.

The science of statistics developed gradually and its field of application widened day by day. Hence it is difficult to give an exact definition of statistics. The definition changed from time to time depending upon its use and application. Numerous definitions have been coined by different people. These definitions reflect the statistical angle and field of activity. A few of these definitions are given below :

The word "statistics" is known to have been used for the first time in "Elements of Universal Erudition" by Baron J.F. von Bielfeld, translated by W. Hooper M.D. (3 vols., London, 1970). Here statistics is defined as :

"The science that teaches us what is the political arrangement of all modern states of the known world".

Webster : "The classified representing the condition of the people in a state, especially those facts which can be stated in numbers or in tables of numbers or in any tabular or classified arrangement".

Horace Secrist : "Statistics is an aggregate of facts affected to a marked extent by the multiplicity of causes, numerically, enumerated or estimated according to reasonable standard of accuracy, collected in a systematic manner for a predetermined purpose and placed in relation to each other".

Prof. A.L. Bowley gave several definitions of statistics as :

(i) The science of counting.
(ii) The science of averages.
(iii) The science of measurement of social phenomena, regarded as a whole in all its manifestations.
(iv) A subject not confined to any one science. A.L. Boddington "Statistics is the science of estimates and probabilities".

Croxton and Cowden : "Statistics may be defined as the collection, presentation, analysis and interpretation of numerical data".

Wallis and Roberts : "Statistics may rightly be regarded as a body of methods for making wise decisions in the face of uncertainty".

R.A. Fisher : "The science of statistics is essentially a branch of applied mathematics and may be regarded as mathematics applied to observational data".

Of all the definitions, the one given by Fisher is considered to be the most exact. Fisher's definition is most exact in the sense that this covers all aspects and fields of statistics.

In view of the latest developments in the field of statistics, it is considered as a science of decision making, under uncertainty, with or without data.

The credit of widening the use and scope of statistics mainly goes to the statisticians of England. The concept and field of statistics in the twentieth century has been changed totally.

Further, the theory of inference, design of experiments and sampling theory were developed which proved landmarks in the development of statistics. Some other contributions further support this point of view. Francis Galton invented the regression theory and pioneered the use of statistical methods in biometry. Karl

Pearson developed the theory of distribution and correlation analysis. His invention of Chi-square test brought statistics to the lime light. W.S. Gosset developed t-test in 1908 and called it the student's t-test. R.A. Fisher did a lot of work in various directions namely, the theory of estimation, the fiducial inference, exact sampling distributions, the theory of design of experiments and testing of hypothesis. Fisher used statistical methods in a number of sciences like biometry, agriculture, genetics, sociology and education. His concepts were fundamental and oriented to the application side. Since then the field of statistics had been widening day by day and is increasingly being used in sociometry, psychometry, biometrics, and technometrics etc.

The statistical methods or techniques are applicable only when some data are available irrespective of the method of data collection. The data can be quantitative as well as qualitative. If the data are qualitative, they are quantified by using techniques like ranking, scoring, scaling or coding etc. The data are collected either by experiments or by survey methods (directly or indirectly) and they are tabulated and analysed statistically. Whatever may be the resulting value(s) obtained from analysis, proper and correct inference have to be drawn from these numerical values. These inferences lead to a final decision.

On the basis of these ideas, we can broadly give the following functions of statistics :

(i) Collection of data
(ii) Tabulation of data
(iii) Analysis of data
(iv) Interpretation of results.

The four functions given above will be described adequately later. Further, their application and utility will obviously be clear from the discussion of the subject matter given in the body of this book.

COLLECTION OF DATA

Once it is decided what type of study is to be made it becomes necessary to collect information about the concerned study, mostly in the form of data. For this, information has to be collected from certain individuals directly or indirectly. Such a technique is known as survey method. These are commonly used in social sciences i.e., the problems relating to sociology, political science, psychology and various economic studies. In surveys, the required information is supplied by the individual under study or is based on measurements of certain units. Generally the respondents or units are selected from a population using some standard sampling techniques. Another way of collection data is by experimentation, i.e., an actual experiment is conducted on certain individuals or units about which the inference is to be drawn. Observations (measurements or counts) are taken on the subject under study. Such experimental studies are common in agriculture, medical science, chemistry, industry etc.

Types of Data

There are two categories of data namely, (i) primary data and (ii) secondary data.

The data, which are collected from the units or individual respondents directly for the purpose of certain study or information, are known as primary data. For instance, an enquiry is made from each tax payer in a city to obtain their opinion about the tax collecting machinery. The data obtained in a census study are termed as primary data. If an experiment is conducted to know the effect of certain fertiliser doses on the yield or the effect of a drug on the patients, the observations taken on each plot or patient constitute the primary data. Hundreds of such examples can be cited.

The data, which had been collected by certain people or agency, and statistically treated and the information contained in it is used again from records, processed and statistically analysed to extract some information for other purpose, is termed as secondary data. For instance, if the data given in different census years is again processed to obtain trends of population growth, professional changes, changes in sex ratio, mortality rate etc., it is termed as secondary data. Usually secondary data is obtained from year books, census reports, survey reports, official records or reported experimental findings. Different organisations and government agencies publish information (data) in the form of reports, periodicals and journals etc. Names of organisations and their publications with relevant details are given in the chapter on Statistical Organisation in India.

PROCESSING OF DATA

Before tabulation of primary data, it should be edited for (i) completeness, (ii) consistency, (iii) accuracy, and (iv) homogeneity.

If the answer to some important question in a schedule or questionnaire is missing, it becomes necessary to contact the informant again and complete the missing information cannot be completed, the schedule or questionnaire should be discarded or revised.

Some information given by the respondent may not be compatible in the sense that an information furnished by the individual either does not justify some other information or is contradictory to earlier one. For example, the total expenditure exceeds the income reported by the respondent, the number of children mentioned is less than total number of sons and daughters, then the respondent should again be contacted to rectify the mistake.

It is of vital importance, if the data are inaccurate, the conclusions drawn from it have no relevance or reliability. By checking the schedules or questionnaire only a little improvement can be made e.g., if the sum of certain figures is wrong, it can be corrected. But if the investigator has either made a wrong report or the respondent has deliberately supplied the wrong information about his income, age or assets etc., editing will be of no use. In recent times, checks have been evolved to attain accuracy e.g., sending supervisors to check the work of investigators or reinvestigating a few respondents after a certain gap of time.

To maintain homogeneity, the information sheets are checked to see whether the unit of information or measurement is the same in all the schedules. For instance, some people might have reported income per month and some annual income. In such a situation it has to be converted to the same unit during editing. It should also be checked whether or not the same information has been supplied for a particular question in all the information sheets. The ambiguity arises due to various interpretations of the same question and should be removed.

Once the primary data have undergone the above four processes it is fit for further analysis.

Large scale data cannot be collected repeatedly because of the paucity of time, money and personnel. Hence the use of secondary data, one should always take care of the following points.

(a) One should see whether the data are suitable for study.
(b) The source of data should also be viewed, keeping in mind whether at any time, it is reliable or not. If there is any doubt about the reliability of data, it should not be used.
(c) It should be noted that the data on a sample, one should see whether the sample is a proper representative of the population.
(d) In case the data are based on a sample, one should see whether the sample is a proper representative of the population.
(e) The primary data has been handled carefully by skilled persons only.

Once the above points are observed in the secondary data, it is ready to be used for further analysis.

Accuracy of measurements depends on the precision required, the tools available for the measurement and the skill of the person undertaking the task. Though absolute accuracy is neither possible nor desired in statistics, a reasonable degree of accuracy is a must. For instance, the monthly income of a person should be rounded to the nearest of ten of rupees, height is measured up to one-tenth of a cm and calculation of age in years and months is fairly accurate.

Sometimes the figures (values) are rounded by reducing one or more decimal places to the unit place or nearest to the ten or hundredth of a number. The universal rules of rounding are, if the decimal place value to be rounded is less than 5, it should be deleted straightaway, and if greater than 5, the preceding number is increased by 1. In case the decimal place value to be rounded off is

5, the rule is to delete it if the preceding number is even and increase the preceding number by one if it is odd.

For instance, 9.74 will be rounded to one decimal place as 9.7 and 9.78 will be rounded as 9.8 Moreover, 9.75 will be rounded as 9.8 and 9.85 will also be rounded as 9.8 according to the rule. The rule is applicable in general for any numerical value.

Before giving absolute and relative error it is worth pointing out that in statistics an error is different from a mistake. The mistake in counting, weighing, measuring or reporting are not errors in statistical sense and should be regarded as mistakes only. Error in statistical sense means the difference between the actual value and the value under consideration, generally an estimated value. An absolute (*A.E.*) is the absolute difference between the actual value (X) and its estimated value (X).

$$A.E. = [X - X] \qquad (1.1)$$

The relative error (R.E.) is the ratio of the absolute error to the actual value, i.e.

$$R.E. = \frac{[X - X]}{X} \qquad (1.2)$$

Very often the relative error the better is given in percent, i.e.

$$R.E. = \frac{[X - X]}{X} \times 100 \qquad (1.2.1)$$

The smaller the relative error the better the result.

Methods of Enquiry

The objective of a study, the population under study, the units or individuals from whom the information is to be collected is determined first. Then keeping in view the purpose and importance of the investigation and types of respondents, the statistician has to choose one of the three methods of enquiry given below.

(1) Personal enquiry method.
(2) Correspondence, i.e. mailed questionnaire method.
(3) Direct observational method.

Before starting the investigation, a question sheet is prepared which is called schedule. The schedule contains all the questions which would extract a complete report from a respondent. Often, the schedule contains the likely answers also. As a precaution in any method a few schedules are tested by filling them before the commencement of the survey. This pre-testing of schedules removes certain discrepancies like ambiguity of the questions and irrelevant questions in the schedule. Such pre-testing is known as a pilot survey.

In this type of enquiry, the investigator contacts the respondent personally and asks him questions given in the schedule one by one and notes down his replies on the schedule. If the unit is a household, the investigator contacts the head of the family to fill up the schedule. Such a practice is called a direct personal interview method.

Sometimes, the information is not collected directly from the respondent but from a third person who is expected to know him well. Such an approach is useful in case where the respondent is expected to hide information about himself. For instance, a person who is abdicated to alcoholic drinks or suffering from some infectious disease may hide correct information about certain things. Hence, an indirect enquiry will give better information. Such a method of enquiry is known as indirect personal enquiry method. The main advantage of this method is that the information is likely to be complete, more correct and some additional information can be retrieved. The person may be persuaded to reply some of the more personal questions too.

When a survey is spread over a vast area and the respondents are educated, a mailed questionnaire method is preferred in comparison to the personal interview method. In this method, a questionnaire is mailed to each and every respondent. They are requested to fill up the questionnaire and send it back. This method is that a large number of respondents do not return the duly filled in questionnaires.

In this approach an investigator stays at the place of survey. He does not make enquiries but notes down the observations himself. For instance, to understand the migratory habits of a variety of birds, the investigator stays at the birds sanctuary and notes down the movements to and fro every day. This method of survey is not used frequently.

Tabulation of Data

The information collected through an enquiry or experiment may be presented in the form of tables of graphs. These tables enable us to come to some conclusion and make certain comparisons. Besides this, they set a stage fit for analysing the data. A table is a systematic arrangement of data in rows and columns, which is easy to understand and makes data fit for further analysis and drawing conclusions. On the other hand, a diagram, a chart or a graph is the visual form for presentation of data. Such a presentation makes it easy for a common man to understand the fluctuations, variations and the existing state of affairs of a phenomenon. The diagrams, charts or graphs may be two or three dimensional. The charts containing pictures with captions are called pictograms.

Analysis of Data

As given by Zacks[1], "one of the most important objectives in the primary stage of statistical analysis is to process the observed data and transform it to a form most suitable for decision making. This primary data processing generally reduces the size of the original sets of sample values to a relatively small number of statistics. It is desired, however, that no information relevant to the decision process will be lost in this primary data reduction".

The measures of central tendency and dispersion like mean, median, mode, mean deviation and standard deviation etc., are parts of analysis of data along with estimation and testing of hypothesis. Besides these, any other statistical tools used or operations done for drawing inference or making decision on the basis of data, are known as analysis of data.

Interpretation of Data

Once the data have been analysed, some numerical value(s) which gives information partly or wholly about the population under study can be achieved. The main job consists of attaching physical meaning and giving interpretation to the numerical results useful in real life. This must be true in its meaning and sense. The quality of interpretation depends more and more on the numerical results obtained out of analysis of data. Also no attempts should be made to draw more inferences than these results are actually liable to.

Especially spuriously calculated values should have no conclusions. For example, if a yearly data for the last decade regarding the production of steel and production of shoes is taken and one finds a significant positive correlation[2] between the production of steel and production of shoes, but it should never be interpreted that production of shoes increase in proportion to the production of steel. These two items are in no way related with regard to their production. Hence such a correlation is spurious and carries no sense.

Limitations of Statistics

1. The main limitation of statistics is that it cannot deal with a single observation or value.
2. Statistical methods are not applicable to studies which measure qualitative characters and cannot be coded in numerical values.
3. The statistical study does not take care of the changes occurring to the individuals. But it does reveal the changes occurring in a mass or group of individuals. For instance, per capita gross national product (GNP) at factor cost in 1970 is Rs. 673.80 and in 1980 it is Rs. 1690.30. On the basis of these figures it can definitely be concluded that per capita GNP has increased by more than Rs. 1000 but it must be kept in mind that it gives no information about an individual.
4. Statistical statements or conclusions are generally not true or applicable to individuals but are applicable to the majority of cases. For instance, the statistical information

says that 90 per cent patients of a particular disease die during operation. If a doctor has performed nine operations so far and all the nine patients have died during operation, he still cannot assure the tenth patient that he will be cured after the operation.

There can be hundreds of situations which can be enunciated under the heading of limitation of statistics. These ideas are of common sense and can be thought of by individuals themselves.

Misuse of Statistics

Some people think that the inferences based on statistical data are very reliable as the numbers cannot be false while others do not trust statistical results at all. According to the non-believers these numerals are the tissues of falsehood as Disraeli said : "There are three degrees of lies, damned lies and statistics". Darrell Huff[3] said, "A well wrapped statistics is better than Hitler's big lie, it misleads, yet it cannot be pinned on you". Another statement by Huff[4] showing his distrust in statistics is "The secret language of statistics, so appealing in a fact minded culture, is employed to sensationalise, inflate, confuse and oversimplify". All this confusion has emerged due to the fact that it is not easy to differentiate between fictitious data and the data collected and tabulated in a well planned manner.

Statistics is misused either deliberately or often due to the lack of knowledge. As defined earlier statistics is a tool, technique or an approach to deal with the data to arrive at correct conclusion or decision. It intelligibly used, it gives wonderfully good results and if misused, it can be disastrous. This can be very well supported by the statement, "Figures won't lie, but liars figure". Another statement in support of the misuse of statistics is, "Statistics is like clay, of which you can make a god or a devil".

The conclusion one comes to it, that at present times, statistics is a highly developed science with a deep rooted mathematical base. It is applicable to a large number of social economic and business phenomena. Statistics is the backbone of industrial research, basic science research and planning.

CLASSIFICATION AND TABULATION OF DATA

The raw data collected through surveys or experiments will be in a haphazard and unsystematic form. Such a data is not appropriately formed to draw right conclusions about the group or population under study. Hence it becomes necessary to arrange or organise data in a form, which is suitable for identifying the number of units belonging to a more classified group, for comparisons and for further statistical treatment or analysis of data etc.

The placement of data in different homogeneous groups, formed on the basis of some characteristics or criteria, is called classification of data. For instance, the people may be divided into different age groups like < 10, 10-20, 20-30, 30-40 etc., or may be classified according to their monthly income like < 500, 500-750, 750-1000, etc. Further, these classified data can be clearly the values or number of units possessing the required characters or belonging to specified classes.

According to L.R. Connor, classification is the process of arranging things (either actually or notionally) in groups or classes according to their resemblance and affinities, and gives expression to the unity of attributes that may subsist amongst the diversity of individuals. Professor A.R. Ilersic has stated that the statistician's first task is to reduce and simplify the details into a form so that the salient features may be brought out, while still facilitating the interpretation of the assembled data. This procedure is known as classifying and tabulating the data.

In short, a table is a systematic arrangement of data in rows and/ or columns. As a matter of fact, the kind of classification or tabulation mostly depends on the type of information required for study and the type of further statistical treatment to be undertaken. Though, there are no hard and fast rules, still some norms can be given for an ideal classification and tabulation of data. Norms for an ideal classification are :

1. The classes should be complete and non-overlapping. It means that each observation or unit must belong to a unique class. For instance, if we classify them as married and unmarried. But there are many who do not belong to either

of these groups namely, divorcees, widows or widowers. Since the number of such people is very small as compared to the number of married and unmarried people, they can be classified as 'others'.

2. Clarity of classes is another important property. It means, classes should be such that one can place a unit or an observation in a class without confusion.
3. One should use standardized classes so that the comparison of results can be possible from time to time. For example, to know the educational development, we should use classifications like illiterate, literate, primary, secondary, graduate, post-graduate and technical. Further, the unit of each class should be the same. The classification of data is generally done on geographical, chronological, qualitative or quantitative basis on the following lines :

   (i) In geographical classification, data is arranged according to places, areas or regions.

   (ii) In chronological classification, data are arranged according to time i.e., weekly, monthly, quarterly, half-yearly, annually, quinquennially, etc.

   (iii) In qualitative classification, the data are arranged according to attributes like sex, marital status, educational standard, stage or intensity of disease etc. It is not essential that a table should have only one attribute. The data can be classified according to any number of attributes. If the data can be classified according to any number of classes is more than two, it is said to be a manifold classification. The readers will come across many examples for these situations in the discussion later.

   (iv) Quantitative classification means arranging data according to certain characteristic that has been measured e.g., according to height, weight or income of persons, vitamin content in a substance etc. In this type of classification, certain classes are formed and the units belonging to these classes are attached to them. One problem that arises is of determining the class intervals. This class interval will depend upon the number of classes which will be arbitrarily decided keeping in view the quantum of data, measurement of characteristics and type of information required. A numerical formula as suggested by H.A. Sturges may be used of classes. The formula is,

$$i = \frac{L-S}{1\ 3.322\ \log 10^{n}} \quad (2.1)$$

where $1+3.322 \log 10^{n} = k$, the number of classes

i = class interval

L= largest observation

S=smallest observation

n=total number of observations

Time Series : Another type of classification is the time series in which data, on the derived values from data for each time period, are arranged chronologically. Extensive examples of the time series data are given in the chapter on Time Series Analysis.

Stem and Leaf Display of Data

Suppose we have a set of n observations $X_1, X_2, \ldots, X_n$. A simple and easy way to represent these data is the stem and leaf display. The methodology of this kind of display is to divide each of the value x_i (i=1m2...,n) into two parts. One part consists of one or more leading digits (highest and next highly placed value digits) as stem and rest of the digits as leaf. The stem values are listed out to the left of vertical line and each leaf value corresponding to a stem is written in the horizontal line to the right of the stem in the order in which they are encountered in passing from one value to the other. For example, we have eight two digit values as 43, 48, 52, 66, 53, 47, 68, 57. The set of observations can be presented in the form of stem and leaf display as given below by taking the digits in the tenth place as stem and in the unit place as leaf.

Stem	Leaf
4	3, 8, 7
5	2, 3, 7
5	6, 8,

Example 2.1 : The data below give the monthly output of coal in India after independence from July 1948 to June 1951.

Monthly Output (Million Tonnes)

22.3,	22.3,	25.6,	24.6,	22.2,	25.7,	25.7,	27.1,	27.0
27.0,	26.0,	23.7,	23.8,	26.3,	27.3,	25.9,	26.0,	28.8,
26.1,	29.4,	28.8,	27.2,	27.1,	27.4,	23.0,	26.0,	28.2,
26.2,	25.6,	28.2,	28.1,	28.9,	29.6,	28.5,	29.7,	26.7

The given data can be presented by stem and leaf display as follows :

The lowest value in the given data is 22.2 and highest value is 29.7. So we can take two digited whole numbers from 22 to 29 as stem and the figures in decimals as leaves. Thus, the stem and leaf display with eight stems is as given below :

Stem	Leaf
22	.3, .3, .2
23	.7, .8, .0
24	.6
25	.6, .7, .7, .9, .6
26	.0, .3, .0, .1, .0, .2, .7
27	.1, .0, .0, .3, .1, .4
28	.8, .8, .2, .2, .1, .9, .5
29	.4, .6, .7

The above display demonstrates they monthly production of coal and makes us to know about the data in a systematic manner.

PRESENTATION OF DATA

Through the data presented in the form of table yields good information to the educated people, they are not always good for the layman as the table cannot be understood easily by them. Moreover, different people may draw different inferences from the tables. Hence the data are presented in the form of diagrams, charts, graphs and pictograms etc., which is good for obtaining visual information. Out of a variety of diagrams, charts and graphs, a suitable one can be chosen keeping in view the type of data at hand, information to be depicted and the kind of recipients. The size of the diagrams and charts should also be according to the place where they are to be displayed.

Generally, the diagrams, charts or graphs are two dimensional though at times they can be three dimensional. In a two-dimensional figure there are always two axes named as *X*-axis or abscissa and *Y*-axis or coordinate. The two axes intersect at a point which is known as the origin and all distances are measured from the origin. The independent variable like the name of place, the time periods or the age etc., are taken on the *X*-axis and the dependent variable like the income, the weight, the marks etc., are taken on the *Y*-axis. The scale is also indicated on these axes.

Before giving the description of diagrams, it is essential to familiarize oneself with certain terms.

Variable : Commonly a factor or character which can take different values is called a variable e.g., the height, length, weight, age, income and expenditure etc.

Random Variable : In mathematical sense, a random variable (*r.v.*) is a real valued function $\{f(x)\}$ defined over a specified range or over a sample space.

Note : Random variable is elaborately and more specifically discussed in Chapter IV.

Continuous Random Variable : A random variable which can take on a continuum of values is called a continuous *r.v.* In this case, the values are taken on a line within the specified range. For instance height, weight etc.

Discrete Random Variable : A random variable which can take a finite or denumerable number of values e.g., the number of students in a class, the number of spots obtained in a throw of dice etc.

Frequency : Number of times a variate value is repeated is called frequency of the variate value e.g., suppose there are seven girl students who have secured 54 marks, 7 is the frequency of 54 marks. If there are 12 people with monthly income of Rs. 500-700, 12 is the frequency of the income group 500-700.

Now some commonly used diagrams, charts and graphs are described here with the aid of actual data.

Frequency Array : If the individual items or values of a variable are given along with their corresponding frequencies, it is called a frequency array.

Example 2.2 : The wages per month and the number of persons in a small scale industry are presented below :

Wages per month (Rs.)	350	490	600	780	800	100
No. of persons	4	5	7	8	4	2

Such a presentation of data is called frequency array.

Frequency Distribution : The premise of data in the form of frequency distribution describes the basic pattern which the data assumes in the mass. Frequency distribution gives a better picture of the pattern of data if the number of items is large enough.

From a frequency array, it is not possible to compare characteristics of different groups. Hence for this, the classes are established to make the series of data more compact and understandable. Class limits can sometimes arbitrarily or by Sturge's formula be delimited. The width of a class, that is the difference between the upper and lower limit of the class is termed class interval. Once the classes are formed, the frequencies for these classes from raw data are expedited with the help of tally marks little slanting vertical strokes. A bunch of four marks is crossed by the fifth to make the counting simpler.

Example 2.3 : In a survey, the age of 52 women at marriage by age parishes was reported as given below.

24,	25,	27,	26,	22,	23,	24,	25,	24,	25,	24,	23,	26,
28,	24,	25,	23,	24,	25,	25,	24,	25,	25,	22,	27,	28,
27,	26,	25,	24,	25,	28,	26,	25,	27,	25,	24,	27,	24,
25,	25,	24,	25,	24,	26,	27,	25,	27,	26,	25,	28,	26.

The data can be presented in the form of frequency distribution with the help of tally marks.

Classification, Tabulation and Graphical Representation

Age in years (i)	Tally marks (ii)	No. of women (iii)
22	11	2
23	111	3
24	~~1111~~ ~~1111~~ 11	12
25	~~1111~~ ~~1111~~ ~~1111~~ 11	17
26	~~1111~~ 11	7
27	~~1111~~ 11	7
28	1111	4
29		0

The distribution constituted by columns (i) and (iii) in the above table is known as frequency distribution. It gives the number of women according to their age at marriage i.e., two women were married at the age of 22 years, three at the age of 23 years, twelve at the age of 24 years and so on.

The frequency distribution has helped to arrange the haphazard data in a systematic manner which is easy to handle for further treatment.

Example 2.4 : The birth weights (kilogram) of 30 children were recorded as follows :

2.0,	2.1,	2.3,	3.0,	3.1,	2.7,	2.8,	3.5,	3.1,	3.7,
4.0,	2.3	3.5,	4.2,	3.7,	3.2,	2.7,	2.5,	2.7,	3.8,
3.1,	3.0,	2.9,	2.8,	2.9,	3.5,	4.1,	3.9,	2.8,	2.2.

Frequency distribution can be formed in the manner described so far, using various class intervals. The width of the classes and the number of classes will be found out by Struge's formula (2.1). The range of data is 2.0 to 4.2 i.e.

$$L = 4.2, \qquad S = 2.0.$$

The class interval

$$i = \frac{4.2 - 2.0}{1 + 3.322 \ \log 10 \ 30} = \frac{2.2}{1 + 3.322 \times 1.4771} = \frac{2.2}{5.91} = 0.37 = 0.4$$

and $K = 5.91 = 6$.

Hence six classes with a width of 0.4 kg. are to be taken in the frequency distribution. The distribution with the help of Tally marks is,

Classes (weight in kg.)	Tally marks	No. of children (Frequency)
2.0-2.4	~~1111~~	5
2.4-2.8	~~1111~~	5
2.8-3.2	~~1111~~ 1111	9
3.2-3.6	1111	4
3.6-4.0	1111	4
4.0-4.4	111	3

Note :

1. The lower limit of a class is included in the class.
2. It is not necessary to choose the smallest the value as the lower limit of the lowest class or the largest value as upper limit of the highest class. One may choose the classes as 1.0-1.4, 1.4-1.8 and so on.

Smoothening of a Grouped Distribution : In case the classes do not constitute the continuous distribution, i.e., the upper limit of the previous class is not the lower limit of the following class, it has to be made continuous. The simple way to do this is to find the difference of the upper limit of the preceding class and lower limit of the following class. Subtract half of the difference from the lower limit of the following class and add the same to its upper limit. Continue this process for all the classes.

Example 2.5 : The table below gives the distribution of the women at the time of marriage in Sri Lanka.

Age groups (Year)	No. of women
15-19	11
20-24	36
25-29	28
30-34	13
35-39	7
40-44	3
45-49	2

The given distribution is not continuous as the upper limit of the preceding class is not the lower limit of the following class. Hence it is smoothened. The difference between 20 and 19 is 1. Therefore, 0.5 is to be subtracted from the lower limit of the classes and 0.5 is to be added to the upper limit of all classes. Since the difference is constant, the same quantity is subtracted and added in all classes.

Thus, the smoothened frequency distribution will be :

Smoothened age groups (Years)	No. of women
14.5-19.5	11
19.5-24.5	36
24.5-29.5	28
29.5-34.5	13
34.5-39.5	7
39.5-44.5	3
44.5-49.5	2

Open End Classes : An open end class is a class lacking one limit. Generally it is the lowest class lacking the lower limit and highest class lacking the upper limit. For instance, in an age group distribution, the lowest class is taken as less than five (< 5) and highest class as more than seventy (> 70). Open end classes make it possible to accommodate values which are at large gaps without increasing the number of consecutive classes. However open end classes should be avoided as far as possible. Open ends create problem in processes like computations and graphical representations.

Cumulative Frequency (cu. fr.) : It is the number of observations less than (more than) or equal to a specified value.

Cumulative Frequency Distribution : It can be formed on "less than" or "more than" basis. In example 2.5, we can either base the distribution on the number of children with a birth weight less than a particular weight or the number of children with a birth weight more than a specified weight. The cumulative frequency distributions for the data given in example 2.5 are presented below.

CUMULATIVE FREQUENCY AND PERCENTAGE DISTRIBUTION

< Type		Percentage	> Type		Percentage
Birth Weight	C.F.		Birth Weight	C.F.	
Less than 2.4	5	17	2.0 or more	30	100
Less than 2.8	10	33	2.4 or more	25	83
Less than 3.2	19	63	2.8 or more	20	67
Less than 3.6	23	77	3.2 or more	11	37
Less than 4.0	27	90	3.6 or more	7	23
Less than 4.4	30	100	4.0 or more	3	10

REFERENCES

1. Zacks, S., *The Theory of Statistical Inference*, John Wiley & Sons, New York, 1971, p. 29.
2. Correlation coefficient has been discussed in Chapter 5.
3. Darrell Huff, how to lie with statistics, p. 9.
4. *Ibid.*, p. 8.
5. The totality of outcomes of a random experiment is called sample space.

2

Measures of Central Tendency

Generally average is of a very great significance as far as comparisons are concerned. It is only for the purpose to find out a single representative value for the whole region or the value around which each distribution tends to have the maximum concentration of observations. This value is the best representative of the whole distribution and the different ways through which it is measured are known as the measures of central tendency or the measures of location.

Mean

Mean or arithmetic mean is the simplest form of various types of averages which is defined as the sum of the items in the data divided by the number of items in the data such as if there are 10 figures of the height of students in inches :

60, 65, 70, 64, 68, 72, 70, 64, 60, 63

It is a general method to add all the values and divided by the number of students, then average height will be obtained which will be known as the mean or arithmetic mean. Thus the total 656 divided by 10 is '65.6', the mean height of the students. It is represented in the following manner :

$$\overline{X} = \frac{X}{n}$$

In case of grouped data the middle values of each class are considered as X and the rest of the methods remain the same.

Suppose the data of ten years of rainfall for a location is given as below. Compute the average rainfall of that location.

Station	A	B	C	D	E	F	G	H	I	J
Rainfall (inches)	25	27	29	40	35	34	33	32	39	38

Solution : Arithmetic Mean

$$AM = \frac{25 + 27 + 29 + 40 + 35 + 34 + 33 + 32 + 39 + 38}{10}$$

$$= \frac{332}{10} = 33.5 \text{ inches}$$

Exercise

A factory has employed 132 workers on varying salary. Calculate the average salary of the workers.

No. of workers	8	12	15	20	22	25	30	Total 132
Salary	50-60	60-70	70-80	80-100	100-120	120-150	150-200	

Classes Salary	No. of workers (f)	Mid value of classes (x)	$u = X - 90$	fu	fx
50-60	8	55	-35	-280	440
60-70	12	65	-25	-300	780
70-80	15	75	-15	-225	1125
80-100	20	90	0	0	1800
100-120	22	110	+20	440	2420
120-150	25	135	+45	1125	3375
150-200	30	175	+85	2550	5250
$n = f =$	132				

The class interval in this exercise is not the same for all the classes, the Arithmetic mean of the above example may be computed by using the following formula.

$$\overline{X} = A + \frac{fu}{n}$$

(1) $AM = \frac{fx}{n} = \frac{15190}{132} = 115.07$

(2) $AM = A + \frac{fu}{n}$ where $A = 90$ $u = X - 90$

$$= 90 + \frac{3310}{192} = 90 + 25.07 = 115.07$$

Exercise

Compute the Arithmetic mean from the following table :

Classes (Rainfall)	No. of days (f)	Mid Value of the classes (X)	$u = X–52.5/5$	fu
30-35	5	32.5	-4	-20
35-40	6	75.5	-3	-18
40-45	11	42.5	-2	-22
45-50	18	47.5	-1	-18
50-55	19	52.5	0	0
55-60	15	57.5	+1	+15
60-65	13	62.5	+2	+26
65-70	1	67.5	+3	+3
70-75	2	72.5	+4	+8
	$N = f = 90$			$fu = -26$

Since the class interval of 5 is common in all the classes, therefore, the following formula is applied.

$$\overline{X} = A + \frac{fu}{n} h = 52.5 = \frac{(-26 \times 5)}{90} = 52.5 - \frac{130}{90}$$

$$52.5 - 1.44 = 51.06$$

Median

In order to get the best representative value when there are some extremely high or low values the arithmetic mean may not be reliable or representative of the average. To measure the central tendency in such a situation the median is considered as the proper method. Median is defined as the middle most value of the observations after arranging them in an ascending or descending order.

For the grouped data if n value of a variable are arranged in ascending or descending order, median is the n + 1/2th, value provided that, n is an odd number. It is taken to be the arithmetic mean of the two middle values if n is an even number.

For grouped data it involves cumulative frequencies also along with the class frequencies and is denoted by :

$$\text{Median} = l + \frac{\frac{n}{2} - C}{f}\, h$$

Where, n is the total number of frequencies, l is the lower class limit in which the median lies.

C is the cumulative frequency of the class proceeding the median class (the class in which the median lies).

f is the class frequency of the median class

h is the class interval of the median class

Exercise

The income of twenty household is given below. Find out their most representative of income.

Since there are some extremely high value the arithmetic mean will give a misleading impression.

Income : 200, 150, 300, 1280, 253, 260, 233, 301, 315, 285, 276, 304, 308, 380, 319, 307, 4989, 100, 403

To compute the median, firstly the data will be arranged in ascending order as below :

100,	150,	200,	233,	260,	276,
285,	300,	301,	304,	305,	307,
308,	315,	319,	380,	403,	1280,
4984.					

The number of observations *n* in this case is even, so there are two middle values (i.e., 301, 304) located. The median would be the mean of these two middle values.

$$\text{Median} = \frac{301 + 304}{2} = 302.5$$

Therefore, the most representative value will be 302.5.

Exercise

The distribution of 2462 towns according to their size of population is given below. What is the most representative size of towns.

Size of towns	Number of towns
<5000	218
5000-10,000	761
10,000-20,000	748
20,000-50,000	484
50,000-100,000	138
> 100,000	113
	2462

In this example arithmetic mean will not be suitable because of the two reasons : (a) the distribution involves extreme values at both ends (b) the middle value of the last class cannot be determined. Thus median is a better measure here and is computed as below :

Size of the Town	Frequencies (No. of towns)	Cumulative frequencies
< 5000	218	518
5000-10,000	761	979
10,000-20,000	748	1727
20,000-50,000	484	2211
50,000-100,000	138	2349
> 100,000	113	2462

$$\frac{n}{2} = \frac{2462}{2} = 1231$$

The median class in which cumulative frequency

$\frac{n}{2}$ = 1231 falls in 10,000 – 20,000.

Therefore,

$$\text{Median} = l + \frac{\frac{n}{2} - C}{f} h$$

Where, l = 10,000 C = 976 mf = 748
h = 20,000 – 10,000 = 10,000

$$= 10{,}000 + \frac{1231 - 979}{748} \times 10{,}000$$

$$= 10{,}000 + \frac{252}{748} \times 10{,}000$$

$$= 10{,}000 + 3368.98$$

$$= 13{,}368.98 \text{ persons (representative)}$$

Mode

The mode value also known as modal value of the distribution is the value which has the maximum frequency. In case of a frequency distribution the mode is computed by the following formula :

$$\text{Mode} = l + \frac{fm - f_1}{2fm - f_1 - f_2} h$$

Where, l is the lower limit of the modal class (i.e., class having maximum frequency), fm is the maximum frequency, f_1 and f_2 are the frequencies of the classes preceding and following the modal class and h is its class interval.

Exercise

Find out the modal size of land holding of a village from the following given data.

Size of holding (acres)	Number of holdings
5-10	20
10-15	35
15-20	150
20-25	70
25-30	44
30-35	38

$$\text{Mode} = l + \frac{fm - f_1}{2fm - f_1 - f_2} h$$

Here $fm = 150$

$$= 15 + \frac{150 - 35}{2 \times 150 - 35 - 70} \times 5$$

$l = 15$, $f_1 = 35$,

$f_2 = 70$ and h = 20 – 15 = 5

$$= 15 + \frac{115}{195} \times 5$$

= 17.94 acres is the representative value

Standard Deviation

Standard deviation is another measure of dispersion in which greater weightage is given to higher deviations about mean while mean deviation gives equal importance to all the deviations about mean or about any other point (median, mode, or any other number). Standard deviation is defined as the positive square root of the average of squares of deviations about mean and is given by

For ungrouped data

$$S.D. \quad = \sqrt{\frac{(X - \overline{X})^2}{n}} \qquad \text{..................(A)}$$

OR

$$S.D. = \sqrt{\frac{X^2}{n} - \frac{(X)^2}{n}} \qquad \text{..................(B)}$$

For grouped data

$$S.D. = \sqrt{\frac{f(X)}{n} - \frac{(\overline{X})^2}{n}} \qquad \text{..................(C)}$$

OR

$$S.D. = \sqrt{\frac{fX^2}{n} - \frac{(fX)^2}{n}} \qquad \text{..................(D)}$$

Formula (B) and (D) are shorter in computation as compared to (A) and (C).

Exercise

Find out the Standard Deviation (S.D.) of the following data.

The rainfall of four months are given as below : 50, 40, 15, and 15 inches. Find out the standard deviation of rainfall.

$$\frac{\left(X - \overline{X}\right)^2}{n} = (50 - 30)^2 + (40 - 30)^2 + (15 - 30)^2 + (15 - 30)^2$$

$$= \frac{20^2 + 10^2 + 15^2 + 15^2}{4}$$

$$= \frac{400 + 100 + 225 + 225}{4}$$

$$= \frac{950}{4} = 237.5$$

Method (A)

$$S.D. = \sqrt{\frac{(X - \bar{X})^2}{n}} = \sqrt{237.5} = 15.41 \quad \text{inches}$$

Method (B)

$$X^2 = 50^2 + 40^2 + 15^2 + 15^2$$
$$= 2500 + 1600 + 225 + 225 = 4550$$
$$X = 50 + 40 + 15 + 15 = 120$$

$$S.D. = \sqrt{\frac{(X)^2}{n} - \frac{(X)^2}{n}} = \sqrt{\frac{4550}{4} - \frac{(120)^2}{4}}$$
$$= \sqrt{1137.5 - 30^2}$$
$$= 15.41 \quad \text{inches}$$

(It is obvious that both the methods have got the same results). We also observed that the standard deviation is higher than mean deviations.

Exercise

Using the method (D) find out the standard deviation of the following data.

No. of persons employed	0-100	100-200	200-300	300-400	400-500	500-600	600-700
Number of units	30	45	50	120	100	112	80
Total no. of units	537						

Method (D)

$$S.D. = \sqrt{\frac{fX^2}{n} - \frac{(fX)^2}{n}}$$

Classes	No. of Units	Mid-Values	fX	fX²
0-100	30	50	1500	75,000
100-200	45	150	6750	1012,500
200-300	50	250	12500	3125,000
300-400	120	350	42000	14700,000
400-500	100	450	45000	20250,000
500-600	112	550	61000	33880,000
600-700	80	650	52000	33800,000
	537		221350	106842500

Using Method (D)

$$S.\,D. = \sqrt{\frac{fX^2}{n} - \frac{(fX)^2}{n}}$$

$$= \sqrt{\frac{106842500}{537} - \frac{(221350)^2}{537}}$$

$$= \sqrt{198961.824 - (412.197)^2}$$

$$= \sqrt{198961.824 - 169906.367}$$

$$= \sqrt{29055.457}$$

$= 170.46$ persons

3

Quantification in Major Fields of Geography

QUANTIFICATION IN GEOMORPHOLOGY

It is a recent phenomenon that geomorphologists have started using the quantitative techniques in analysing the various geomorphic features of the surface of the earth. The analysis was largely related to slope and fluvial is now known as morphometric techniques. For this purpose topographical sheets of various scale was being used for various purposes such as :

(1) River ordering
(2) Measuring the maturity of the rivers
(3) Gradient of the rivers
(4) Intensity and density of the rivers
(5) Measuring the tortuosity of the rivers
(6) Slope analysis : degree of slope instead of elevations, for the purposes of spread of transport network, extension of cultivated area and the urbanization etc.

For the above mentioned analysis only a single source with limited data is available, i.e., topographical sheets with the information of rivers and contour interval.

For each type of analysis various methods and techniques have been used, many of them much improved with the change of time. In this regard Miller[1] has stated that "Much can be learnt from inspection of the map, but when its possibilities by this method have been exhausted there still remains infinitely more that can be extracted by quantitative analysis and rearrangement of the data". The significance and importance of quantification is quite evident from the writings of Goldberg such as "the study of Geomorphometry puts us on the threshold of an even more precise age than before[2]".

It has also been stated that geomorphological science suffers from a lack of quantitative exactness even keeping in view the present state of development in this field[3]. The statement of the following researchers reveals the importance of quantitative method in Geomorphology such as :

G. H. Dury (1952) In geomorphological research there can be no substitute for lengthy and painstaking field work. A multitude of features of the highest significance will appear only on those maps drawn by the field geomorphologist.

Bakker, J. P.; Strahler. A. N. (1956) Full use of the modern morphometric statistical method for slope development can only be made if it is based on field work.

Bauling, H. (1950) There remains to be proved that mathematics has ever revealed an actual relationship in geomorphology that has not been discovered without its aid.

Quam, L. O. (1950) : Efforts towards a more quantitative analysis of landforms is highly commendable, and it is hoped that through investigation of this sort our science may become more objective. Is there not a danger, however, the use of mathematical formulae and statistical analysis give a false impression of objectivity to sample measurements? Are we not in danger of giving an impression of accuracy which is not warranted?

A survey by J. L. Clarke and K. Orrell assesses the reliability of a number of the morphometric methods and stresses their inherent limitations. Now, it is a well established fact that morphometric methods can yield precise, accurate and objective results, superior to general descriptive and deductive method used in the field.

The analysis of slope or gradient of river and its representation on a map has been the subject of much research throughout the world. The calculation of average gradient, either along the steepest slope or along a road is a simple matter. But to work out some representation of average slope, particularly in an area of complicated relief, and to express this on a map to provide a clear picture which may help the geomorphologists to make important deductions, is a much more complicated affair[4].

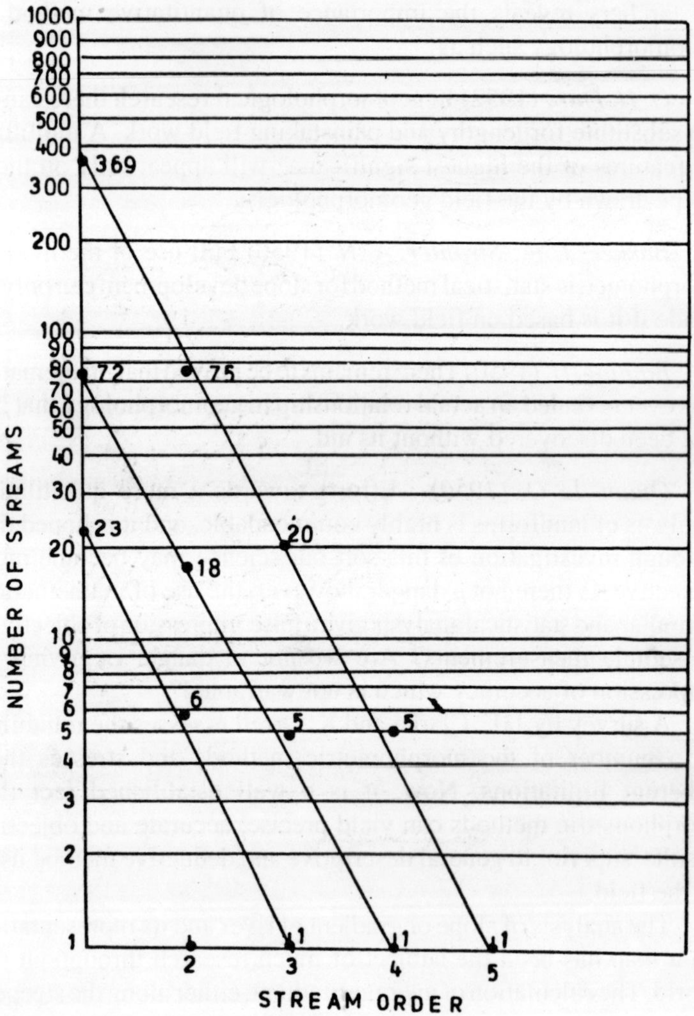

FIG. 1 : River Ordering

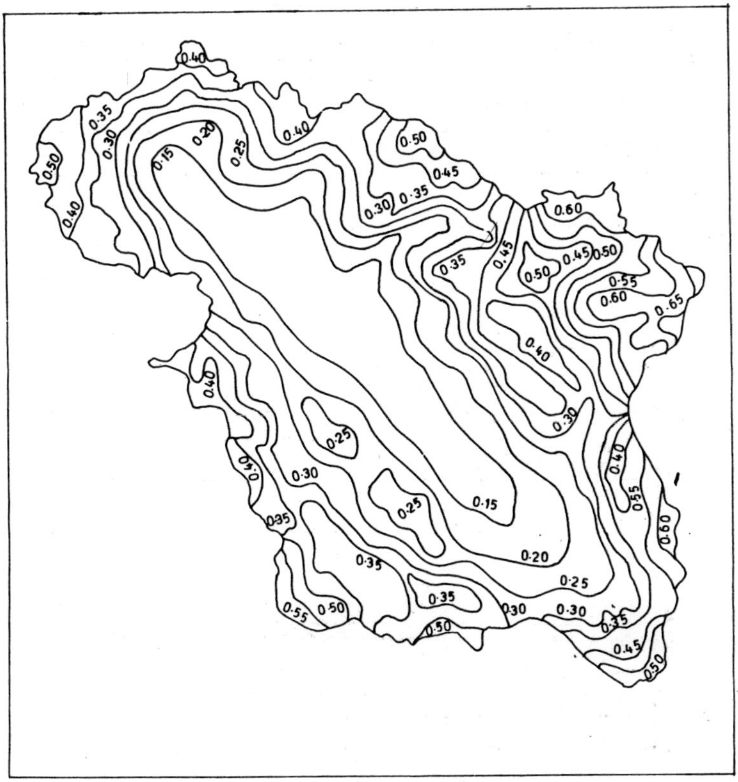

FIG. 2 : Isotangents

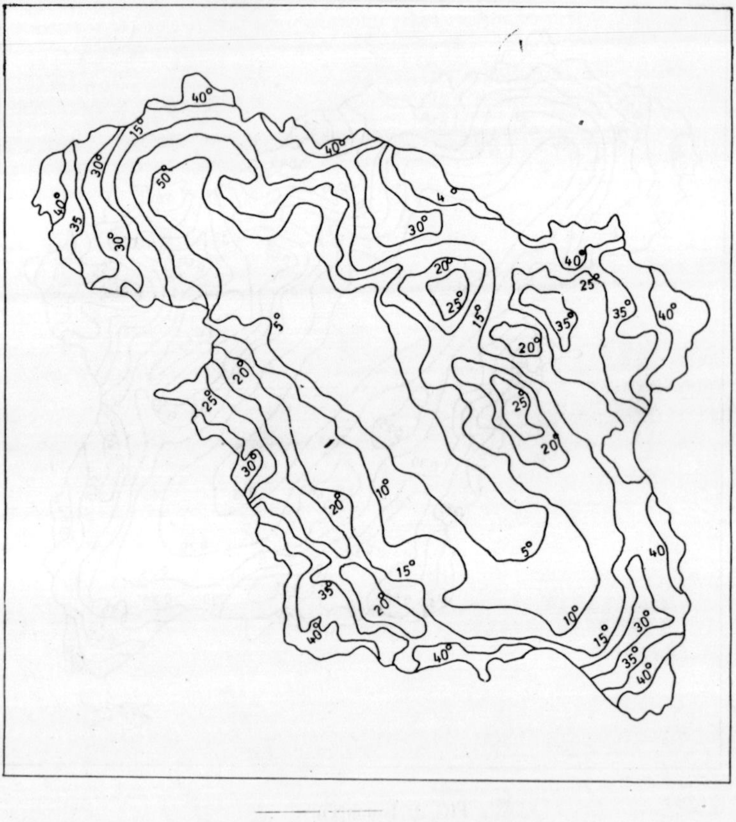

FIG. 3 : Average slope in Degree

Distribution of Slope Categories

Slope (degrees)	Area covered	% of the total Area	Cumulative frequency Sq. km	%
<5°	2118.3	17.2	2118.3	17.2
5°-10°	2448.8	19.8	4567.1	37.0
10°-15°	2276.6	18.4	6843.7	55.0
15°-20°	1521.1	12.3	8368.8	67.9
20°-25°	1211.4	9.8	9576.2	77.7
25°-30°	818.9	6.6	10395.1	84.3
30°-35°	618.1	5.0	11013.2	89.3
35°-40°	520.1	4.2	11533.4	93.5
> 40°	828.6	6.7	12362.0	100.0
	12362	100.0	—	—

Formula:

$$(O)\ \text{Slope (Tangent Value)} = \frac{\text{Average Number of Contour Crossing} \times \text{Contour Interval}}{3361}$$

MORPHOMETRIC ANALYSIS

(A) Slope Analysis : Most of the methods of morphometric analysis are dependent on statistical information available in toposheets such as height of a contour, contour interval, number of contour in an area, distance of contour and general elevations. On the basis of this information various methods and techniques for various purposes have been propounded, such as the gradient of the slope, degree of slope etc. The analysis of river ordering, gradient of the river etc. is also based on the contour and elevations.

Gradient of Slope : For the purposes of transportation network and to measure the intensity of erosion it is essential to find out the exact steepness of the slope.

The distance between the two projected heights is known as the Horizontal Equivalent (H.E.) while the difference in vertical height between the two points is known as the vertical interval (V.I.). This if the H.E. is 100 metres and the vertical interval (V.I.) is 10 metres, the gradient will be 10 :100 or one in 10.

Slope Analysis

For the first time the average slope determination was used by S. Finsterwalder and K. Peucker in the year 1890. Later on this method was considered most laborious and complicated in calculation because it was very difficult to calculate the length of the contours.

$$\text{sine of degrees of Average slope} = \frac{\text{Total length of contours} \times \text{Contour Internal}}{\text{Total area}}$$

In the year 1930 C.K. Wentworth[5] introduced a simply field method for the determination of average slope. The result give the average slope in terms of the tangent of the average angle of slope which can be converted into degree with tangent tables.

$$\frac{\text{Average Number of Contours crossing per Mile} \times \text{Contour Interval}}{3361 \text{ (constant)}}$$

There are various other methods introduced such as :

(a) G.H. Smith Method of Slope Analysis (1935). He established the relationship between the highest and the lowest points in any particular area. He used the term relative relief or local relief.

(b) Raisz and Henry Method of Average Slope Determination (1937). They applied the Smith method but the result in the case of New England was not satisfactory because of the narrow valleys cut deeply into the peneplain.

(c) A.H. Robinsons Method of Slope Analysis (1948). The method derived to produce a quantitatively accurate relief map from areal slope data.

(d) A. N. Strahler's Method of Slope Analysis (1956). Among all these methods the C.K. Wentworth method is considered most useful and applicable as well as very easy to calculate. Most of the methods are largely dependent on this method, therefore, this method is considered as a base of other methods.

Drainage Characteristics of the Rivers

Drainage Basin	Area in sq. km	Drainage Length km	Number of Streams NU	Density $\sum LK/AK$	Texture NU/AK
Liddar	1243	679	281	0.54	0.23
Harwan	395	144	102	0.36	0.23
Sind	1556	766	342	0.43	0.22
Pohru	1936	1130	470	0.58	0.24
Vishav	828	448	163	0.54	0.19
Rembiara	751	373	130	0.48	0.17

1. Drainage Density : $D = \sum \frac{LK}{AK}$

2. Bifurcation Ratio : $RB = \frac{NU}{NU+1}$

Jhelum River Drainage Basin Characteristics

Stream Order	Number of Stream Segments	Bifurcation Ratio	Length of segments in each order and as % of the cumulative: Length km	%	Mean Length of Segments (km)
1	2101		3848	58.02	1.83
2	471	-4.46	1324	19.96	2.81
3	91	-5.18	687	10.35	7.55
4	31	-2.94	468	7.05	15.10
5	8	-3.88	177	2.66	22.12
6	1	-8.00	128	1.93	128.00
	Total	Average	Cumulative Length		
	2703	4.89	6632.00		

QUANTIFICATION IN AGRICULTURAL GEOGRAPHY

1. Land Capability Classification

Land classification is one of the most important prerequisite for a rational landuse planning. It gives us an idea of the productivity as well as the potential productivity of the various land types. This could be taken as base and the selection and combination of crops to be grown should be made in such a way so as to yield maximum results from the varying land type. It is a well known fact that the selection of crops which are to be raised depends on a number of geographical factors including the quality of land.

Landuse planning largely rests on the land of special qualities for special uses. Land capability classification has to do lot with the suitability of land for specific purpose, i.e., for agriculture forestry and grassland and others. It was realised that it is not possible to quantify the land capability data. In the initial stages the land capability analysis is done on the basis of the fertility and productivity of the land and the classification was as follows :

1. *A.* (Good quality land)
2. *B.1*
   B.2 (medium quality land)
3. *C* (Poor quality land)

All the land unproductive were classified under "C" category such as wasteland, saline and alkaline land, waterlogged land and ravine lands etc.

In the year 1955 Bennet. H. H. introduced a classification and divided the land into two basic categories, i.e., land suitable for cultivation. The land suitable for cultivation was divided into four categories and land not suitable for cultivation into four categories. Hence the classification was based on eight types depending on the fertility and uses of the land.

Various attempts were also made to quantify the land capability on the basis of the geographical conditions and human efforts to use the land for various purposes with various means of improvements and controlling the environment. The fertility index of the soil was calculated and other nutrients were also given weightage and the method was introduced as follows :

$$\text{Land Capability} = \frac{\text{Geographical or Physical factors}}{\text{Human Factors}}$$

Land resources for agricultural purpose is related to the combination of various aspects such as soil moisture content in the soil storiness, coarseness and texture of the soil, slope degree etc. While the human factors are associated with the various inputs and management requirement. The value obtained of both the factors are used in measuring the land capability. The selected area is divided into various smaller units and under each unit the value is plotted to get the variations in land Capability Classification.

Measurement of Agricultural Efficiency

Various methods have been evolved to measure the agricultural efficiency considering the various factors such as climate, soil, type of farming, size of the holdings, crop rotation, irrigation and mechanization and human offer which have a combined effect on productivity of a region.

However, on the basis of various variable available and that are needed for such type of study the geographers and agricultural economists have tried their best to evolve various appropriate methods to measure the agricultural efficiency. The methods adopted are based on :

(a) Output per Unit area
(b) Output-input ratio
(c) Output per unit of labour applied
(d) Output in terms of value
(e) Output expressed in terms of per head production.

For the study of Chinese agricultural J. L. Buck (1937) evolved a method as output expressed in terms of grain-equivalent. He has taken all foodgrains to be equivalent food values and basing his unit on one kilogram of whatever kind of grain was predominant in the locality. He converted all other products into grain equivalents on the basis of the amount of the most usually consumed grain of the locality that the product would buy.

A slight modification over the Bucks method was introduced by Vries. He expressed the output in terms of milled rice equivalents per head of population because all the foodgrains converted as equal to rice equivalent according to the local market price of various grains. Buck and Vries seems to have realised the fact, that is, a subsistence agricultural area like China, productivity and value of production expressed in terms of money were not very meaningful. Buck has considered all grains as equal, Vries converted kind of grains into rice equivalent according to the local market price of each grain. This method will not be applicable especially in the multi-regions as there may not be single diet much more prevalent in the dietary habit of the people. A further modification of the procedure has been introduced by the Clark and Haswell who expressed the output from agricultural in terms of kilogram of wheat equivalent per person. For this purpose they employed a scale based on the weighing system used by the Food and Agricultural Organisation of the United Nations for constructing the international index numbers of agricultural production. The results obtained provide an interesting comparison between countries of Africa, Asia and Latin America. The value of a kilogram wheat equivalent per person range from 332 in Lebanon to 529 in Egypt and 1604 in Cuba.

Kendall made an approach to this subject, tracking it as a mathematical problem and devised a system of 'Ranking Coefficients'. This method is based on the acre-yield of crops as output per unit area. He took the area-yield of 10 leading crops for 48 administrative countries and for each of 10 crops he ranked the countries according to yield. The sum of ranks occupied by a country was divided by the number of crops to obtain the average rank of the country. Prof. Dudley Stamp applying his technique on an international basis, selected a number of counties as well as same major crops, the choice of crops depending upon the fact that each one of them was grown in all the counties selected. He placed the counties in the order of output per acre of each crop. The places occupied by each county in respect to the selected crop were then averaged and from these averages the ranking co-efficient of agricultural efficiency of each country was obtained. In India the method was applied for the first time by Shafi who considered the acre-yield of eight crops to measure agricultural efficiency in U. P.

This method seems to have a vital and inherent weakness which makes it somewhat insensitive as a measure of agricultural efficiency. This weakness arises from the neglect of the areal strength of crops for which acre-yields are taken to calculate the ranking coefficient. For instance in Uttar Pradesh Deoria occupies first rank in Bajra on the basis of acre yield though only a fraction of one per cent of crop-land is devoted to it. Obviously, a crop which occupies a negligible portion of the crop land would contribute almost nothing to agricultural efficiency of the areal unit even though it may have a very high acre yield.

Desh Pande and Sapare in order to remove the above noted weakness in Kendall's ranking co-efficient, have modified it by taking weighted average of the ranks (instead of simple average of Kendall), the weights of rank of various crops being proportional to the percentage of crop-land under respective crops. For example, district *A* has rank 4 on the basis of Jowar acre yield, and rank 8 on the basis of gram acre-yield. Also assume that in district *A*, rice occupies 30 per cent of the crop-land Jowar and gram occupy 50 per cent and 10 per cent of the crop-land respectively (these percentages are taken as weights). The weighted average of ranks would be (4× 30) + (16× 50) + (5 × 10) =990 divided by the sum of weights, that is 30 + 50 + 10 = 90 and would equal to 11. In contrast the simple average of ranks according to Kendall would be 4+16+7=27 divided by the number of crops, i.e., 3 and its value would be 9. This method, however, does not make the measure more sensitive than Kendall's ranking coefficient, because the magnitude of difference in acre-yields is concealed by this weakness. A more logical refinement would perhaps be to take the average rank of weighted acre-yields on the assumption that the different crops contribute to the agricultural efficiency according to their acre-yields but proportionate to their share of the total cropland. Therefore, on the basis of the above cited shortcomings of the various methods a new measure introduced by S.S. Bhatra has been considered suitable to measure the agricultural efficiency. This method is based on the weightage of acre-yield in respect of the proportionate of each crop to the total cropped acreage of any aerial unit. This method is as follows :

$$Ii = \frac{\sum_{J}^{P} = E_{ij}\ C_{ij}}{\sum_{J}^{P} = 1\ C_{ij}}$$

$$Eij = \frac{Yij \times 100}{Y_j}$$

Where I_i = Index of crop productivity in *i*th district.
Y_{ij} = Yield per acre of *j*th crop in *i*th district.
Y_j = Yield per acre of *j*th crop in the Region
C_{ij} = Proportionate share of land under *j*th crop in the total cropped area of the *i*th district.

After analysing the agricultural data (in the form of production and acreage of each crop) of any region according to the method explained above, the index value can be obtained and classified into groups by using the index value for the region (i.e., 100) as the dividing line. The index value corresponding to the various degrees of efficiency can be categorised in the following manner.

I. 1. Very low degree of efficiency,
2. Low degree of efficiency

II. 3. Medium degree of efficiency

III. 4. High degree of efficiency
5. Very high degree of efficiency

These categories are not static, it can be changed after a point of time with the change of various factors in agricultural operations. As it is a wide accepted fact that agricultural efficiency is a function of physical, socio-economic and technical organizational factors.

The technique presented above has potentialities of wide application and appears to have immense value in the developing countries. Therefore, it opens up a further line of investigation and perhaps indicates the need of provision of improved seeds and fertilizers and with due attention to better tilling practices, to a balanced rotation of water supply, the level of agricultural efficiency could be considerably increased. If only the agricultural efficiency of the other districts of India is brought to the level of that of the other high efficient districts, the food production of the country will be substantially augmented and this will help in the solution of the food problem of the nation. The study of agricultural efficiency is of particular interest in developing regions like India where there is little scope of further extension of agricultural lands and the only way to meet the increasing pressure of population seems to be the improvement in agricultural efficiency.

INDICES AND GROWTH RATES IN AGRICULTURAL GEOGRAPHY

The purpose of using these devices of index numbers and growth rates is to reduce cumbersome large figures to a form that can be more readily comprehended exposing the relationship more clearly.

(1) Index Numbers

AC = Index of Area under crops

$$= \frac{\sum a_{ij}}{\sum a_{io}} \times 100 \qquad \text{.............(1)}$$

A_p = Index of Agricultural production

$$= \frac{\sum P_{ij} \times P_i}{\sum P_{io} \times P_i} \times 100 \qquad \text{.............(2)}$$

A_{py} = Index of agricultural productivity

$$= \frac{\text{Index of agricultural production}}{\text{Index of area under crops}} \times 100 \quad \text{............(3)}$$

Where

a_{io} = Area under the *i*th crop during the base period
a_{ij} = Area under the *i*th crop during the *j*th year
p_{io} = Production of the *i*th crop during the base period
p_{ij} = Production of the *i*th crop during *j*th year
p_i = Weighted price per unit of the *i*th crop

(2) Linear Growth Rates

For linear growth rates time should be taken as the independent variable and the index numbers of area/ production/ productivity as the dependent variable, a linear equation of the following can be fitted—

$$Y = a + bx$$

Where Y = Index number of Area/ production/productivity.
X = Time

The value of *b* in the above equation represents the slope of line and gives an estimate of the constant amount of increase/decrease per annum, i.e., the rate of change of *Y* with respect to *X*. The constant *a* is *Y* intercept . These are given by :

$$a = \frac{\sum Y - b\sum X}{N}$$

$$\text{and } b = \frac{\sum XY - \frac{\left(\sum X\right)\left(\sum Y\right)}{N}}{\frac{\left(\sum X\right)^2}{\left(\sum X\right)^2 N}}$$

(3) Compound Growth Rates

With the above noted assumptions the following exponential curve can be fitted—

$$Y = abx$$
$$\text{Where } A = \log a$$
$$B = \log b$$

The normal equation are:

$$\sum \log Y = NA + B \sum X$$

$$\sum X \log Y = A \sum X + B \sum X^2$$

These equations can be solved for *A* and *B* then by taking antilog *of a* and *b* can be found out. For example (table)

Distt.	1962			1971			1%Annual Comp. Rate		
	Area	Output	yield	Area	Output	Yield	Area	Output	Yield
A	202918	215733	1063	198566	237283	1195	-0.26	1.20	1.47
B	227522	242864	1067	245016	260942	1065	0.93	0.90	-0.02
C	285583	225928	791	343473	284012	827	2.33	2.90	0.55
D	394770	295960	750	442150	362032	819	1.43	2.55	1.11

(4) Test of Goodness of Fit

This growth rate can be tested through chi-square (χ^2) and their degrees of freedom. Chi-square test is used to know the "goodness of fit" or in other words, to determine whether the deviations of the observed data from the expected data are small or large enough to accept or reject the fitted curve. Each curve is tested at 5 per cent and one per cent level of significance. Wherever the observed value of χ^2 exceeded the table of χ^2 005/ χ^2 001, the difference between the observed and expected data will be considered highly significant and the fit will be considered poor. In other words, if the observed χ^2 is less than or equal to the table value of Chi-square we accept the fit.

COMBINATIONAL ANALYSIS IN AGRICULTURAL GEOGRAPHY

Recently the Combinational Analysis in geographical studies has increased greatly and the importance of this type of study is increasing day by day. Such studies are fruitful in many ways, firstly, it provides an adequate understanding of an individual crop geography, secondly, crop-combination is in itself an interactive reality that demands definition and distributional analysis, and lastly, such regions are essential for the construction of still more complex structure of valid agricultural regions. Moreover, crop-combination regions may be used to evaluate if a particular combination would lead to dietary adequacy or inadequacy in the essentially agricultural landscape.

The study of ranking method reveals that in order to derive primary crop-combination regions, it is not only that rank which counts but also the actual differences in the extent of each crop.

It is better to apply some mathematical procedure for a more objective groupings in spite of arbitrary gradation of percentages. A single scale of gradation was introduced by Johnson for forming crop-association regions in East Pakistan (now Bangladesh). On the basis of the scale of level of importance for each of the five crops calculated from the mean or average for East Pakistan as a whole, areas of very high (VH), High (H), medium (M), Low (L) and very low (VL) importance were demarcated and then grouped into six regions. Johnsons method is very simple. The scale of gradation was calculated only once and the regions are formed by constantly referring to this scale. Although it may be suitable in case of East Pakistan where only five crops are grown and where there is no marked variations in the patterns of crop land-use. The method, however, may not be suitable for the regions where a variety of crops with varying level of importance are produced.

A number of statistical procedures have so far been introduced to demarcate the crop-combination region or set of elements that play significant role in any particular system. A most valuable statistical formula of crop-combination analysis was advanced by J.C. Weaver in 1954 in his study on the crop-combination regions in the Middle West. For determining combination having minimum deviation Weaver calculated deviation of real percentages for all the possible combination in the unit considered against a theoretical standard which was obtained by dividing 100 per cent in one crop, 50 per cent in each of two crop-association, 33.3 per cent in each of three crop-association regions and so on. Weaver formula is as such :

$$\theta = \sqrt{\frac{d^2}{n}}$$

θ = value of the crop-combination

d = the difference between the actual crop percentage in a given unit and appropriate percentage in the theoretical curve.

n = is the number of crops in a given combination.

For example in a unit *X* the percentage of the various crops are as follows :

Wheat 35 per cent, Barley 24 per cent, Peas 15 per cent, Rice 10 per cent, gram 8 per cent, Maize 4 per cent, oil seeds 4 per cent,

1. Monoculture $= \frac{(100-35)^2}{1} = 4225.00$

2. Crop-combination $= \frac{(50-35)^2 + (50-24)^2}{2} = 450.5$

3. Crop-combination $= (33.3-35)^2 + (33.3-24)^2 + (33.3-15)^2 = 14126$

4. Crop combination =

   $= (25\text{-}35)^2 + (25\text{-}24)^2 + (25\text{-}15)^2 + (25\text{-}10)^2 = 106.5$

5. Crop-combination = 99.00

6. Crop-combination = 138.80

7. Crop-combination = 113.20

(Since the variance for 5 crop-combination 99.0 is the smallest in the *X* unit is designated with a combination of five crops i.e., WBPRG).

Weaver's method to determine the minimum deviation appear to be quite simple, but, in practice, it requires much calculation.

An improvement to that of Weaver's the Doi method is more simple, by substituting $\sum \frac{d^2}{n}$ with d^2 or the sum of the squared differences.

Thus the combination having the smallest d^2 will be the combination formula by the major crops only Doi has been defined in detail in his statistical procedure of his paper, The Industrial Structure of Japanese prefecture. His modified formula is that of Weaver which is as follows :

		d^2		
Monoculture	=	$(100 - 35)^2$...............		= 4225
2. Crop Comb.	=	$(50 - 35)^2$ +	$(50-24)^2$	= 901
3. " "	=	" " "	" "	= 424
4. " "	=	" " "	" "	= 426
5. " "	=	" " "	" "	= 495
6. " "	=	" " "	" "	= 833
7. " "	=	" " "	" "	= 792

Since the variance for 3- crop-combination (424) is smallest, the country is designated with a combination of the three crop, i.e., WBP.

On deriving crop-combination for a unit *X* on Doi method, it was found that result are more realistic in comparison to those worked out by other methods. It is easy in calculation.

For delineating crop-combination regions statistical procedures were applied by many but they suffer due to some difficulties. Pownwall method in the functional classification of towns in New Zealand was also introduced and was also improved by Nelson. But the Crop-Combinations based on the simple deviation method are more objective, well defined and realistic than those based on percentages gradation. Nelson's ratings, involve a lot of time and labour in calculating standard deviation for each. Rafiullah introduced a new deviation which, although suffering from defects of laborious calculations, is quite suitable for delineating the crop-combination. The formula is as such.

$$8\sqrt{\frac{Dp^2 - D^2n}{N^2}}$$

Where θ is the deviation

Dp = Positive difference

Dn = Negative difference form the medial value of theoretical curve value of the combination

N = is the number of functions in the combination.

Since only relative rank of the deviation is needed the root sign may be ignored to save calculations and the formula may be used as :

$$8 = \frac{\sum D^2p \sum D^2n}{N^2}$$

Unlike Weaver the difference of actual values is calculated from the middle value of the theoretical standard and the maximum positive deviation gives the critical combination. A comparison of the crop-combination based on Weaver's and that of Rafiullah's will indicate that the latter are either identical or include a lesser number of crops in the combination than the former. The difference is introduced as also pointed out by Rafiullah in only those cases where the difference among the actual values is large. In such cases Weaver's method fails to delineate a precise combination and includes all the crops occupying as much as one per cent of the total.

The point :

(A) *Wg* 20.9, *W.*12.8, *M.*7.5, *Mp* 7.2, *Ok* 6.9, *G.* 6.9, *Mz* 6.7, *R.*6.7, *P.*5.2, *B.*3.5, *L.*3.0, *mp* 2.9, *Be* 2.2, *m* 2.1, *Wb* 1.5

(B) *W* 19.3, *M* 18.4, *Mp* 10.0, *R* 8.5, *Ok* 7.3, *G* 5.5, *Pk* 5.3, *Of* 5.0, *m* 4.0, *B* 3.6, *Mz* 3.6, *WB* 3.1, *L* 1.9, *Be* 1.2, *Wg* 1.2, *S* 1.07.

Number of Crops in the Combinations

Unit	Power Walls Method	Weaver's Method	Rafiullah Method	Doi Method
A	8	14	6	9
B	5	16	6	8

QUANTIFICATION IN DEMOGRAPHY

The numerical portrayal of a human population is known as Demography. The population is viewed as an aggregate of persons represented by certain types of statistics. Demography is concerned with the behaviour of the aggregate and not with the behaviour of individuals. Demographic structure of a country or region plays a very important role in the development processes. Demography covers the following aspects of human population :

1. Age structure : As it affects in various ways in planning the region because it covers the percentage share of workforce from the total population. Dependency ratio is also calculated on the basis of the age structure as higher dependency ratio is related with low characteristics of economic population. A country not well prepared with the emphasis on economic development and opening the various job opportunities if the percentage share of lower age group population is very high, otherwise a country may face the situation of political chaos.

2. *Sex Ratio :* Among the various vital statistics sex ratio plays a very vital role in understanding the future trend of population growth and quality of workforce available in a country.

3. Caste Structure : No doubt that the study of caste in India is important but not in all the countries of the world such as in European countries. In India castes play a very significant role in socio-political economic scene. It is also said that different castes have different aptitude for agricultural operation hence affecting the performance of agriculture. It largely affects the election scenario in India which has the largest democracy in the world.

4. Literacy and Educational Level : The study of literacy and educational level both are different to each other as a country may be having higher literacy rate but low level of education can affect the spread of education needed for the balanced development of a region. It is considered as one of the most important demographic aspects.

5. Workers and Non-Working Population or Workforce and Working Population : The data base related to this problem is very important for the manpower planning. The non-national working force may not be reliable or dependable. If a country is facing the manpower problem of different level skilled or unskilled the developmental objectives of that country may suffer.

6. Economic Characteristics of Population : Population involved in different economic activities such as primary, secondary and tertiary are used to measure the development level of a country and it is also that economic activities are the best indicators to measure the stages of economic development of a region.

7. Income Patterns : To find out the regional inequality in income distribution data of income of various nature is important without which the planning priorities may not be decided.

8. Marital Status : Marital status and age of marriage affect the population growth of a nation. How far the laws imposed regarding this problem is being operated is also matter of concern to bring a nation on normal level.

9. Immigration and Emigration : Unplanned migration from rural to urban has become a major problem in India which is threatening the balanced development of an urban centre. The statistics related to it if not properly collected may lead to a faulty results for urban planning.

DEFINITION OF VITAL STATISTICS

Generally vital statistics is defined in two ways :

1. It refers to all types of population statistics collected by whatever mode.
2. It refers only to the statistics derived from the registration of births, deaths and marriages. According to Benjamin "Vital statistics are Conventional numerical records of marriages, births, sickness and deals by which the health and growth of community may be studied".

SOURCES OF VITAL STATISTICAL DATA

There are various methods to collect the vital statistics data.

1. Census enumeration.
2. Survey method.
3. Registration method.
4. Sample registration system.
5. Analytical method.

Among all these methods, the analytical method is most appropriate and reliable. It is generally not possible to conduct surveys to assess the population at any period in between two census years. The population estimates at a given time can be obtained without *ad hoc* surveys by mathematical methods. The estimates are based on assumption that the population grows at a constant rate during the intercensal years. For example, given the birth and death rates, population growth may be estimated for an intercensal period. The number of births or deaths at a given time in a population sub-group can also be estimated. Some of the methods are as follows :

1. Estimation of population in a given intercensal year can be done by the formula,

$$\hat{P}_t = P_o + \frac{n}{N}(P_1 - P_o)$$

where P_t—Estimated population at some censal year t.

P_o — Population in the previous census.

P_1 — Population in the succeeding census.

N— Number of years between the census.

n — Number of years between the given year and the previous census year.

It is a linear interpolation formula and yields good estimates provided population changes occur at a constant rate all over the intercensal years.

Example

The population of a town in the year 1971 and 1981 are 745876 and 1250532 respectively. The population estimate for the year 1975 is obtained as :

$$P_1 = 1250532,\ P_o = 745876,\ N = 10,\ n = 4$$

$$\text{Thus } \hat{P}_t = 745876 + \frac{4}{10}(1250532 - 745876)$$

$$= 947738$$

2. Population of any intercensal year can be estimated provided accurate records regarding births, deaths and migration are available. The formula is :

$$\hat{P}t = \hat{P}o + (B - D) + (I - E)$$

Where, $\hat{P}t$ and $\hat{P}o$ are the same

B—Total number of births during the period 0 to t.

D—Total number of deaths during the period 0 to t.

I—Total number of immigrants during the period 0 to t.

E—Total number of emigrants during the period 0 to t.

MEASUREMENT OF POPULATION GROWTH

The growth of population of a region is measured through the following methods :

(1) Crude Rate of Natural Increase.
(2) Vital Index.
(3) Gross Reproduction Rate
(4) Net Reproduction Rate
(5) Replacement Index.

CRUDE RATE OF NATURAL INCREASE

This is considered the most simple method of population growth and can easily be obtained by subtracting the crude death rate from the crude birth rate (CBR—CDR). In case the total number of births and deaths of a region in any year is known then the annual crude rate of natural increase can be calculated by the formula :

$$\text{Annual crude rate of natural increase} = \frac{\text{Total birth} - \text{Total death}}{\text{Mid year population}} \times 1000$$

For example

Mid year population	=	3,52,72,000
Total Births	=	187236
Total Deaths	=	72299

The annual crude rate of natural increase by the formula :

$$= \frac{187236 - 72299}{35272000} \times 1000$$

$$= \frac{114937}{35272000} \times 1000$$

= 3.26 per thousand persons.

Vital Index

This is an index which takes into namely, births and deaths. For a specific period and region, the vital index is given by the formula :

$$\text{Vital Index} = \frac{\text{Total births}}{\text{Total Deaths}}$$

The vital Index may be equal to 1 or less than 1 or greater than 1.The value 1 indicates stagnation in population growth. A value greater than 01 throws light on the expected increase in population whereas a value less than 1 is indicative of a decline in the population.

Gross Reproduction Rate : Fertility rates include the birth of children of both the sexes. But the population growth depends mainly on the birth of female children who are the future methods. Hence population growth is mainly a function of the fertility rate, restricted to female children. The demographic year book published by the United Nations, 1954, has defined this term as "The grows reproduction rate indicates the average number of daughters who would be born to a group of girls beginning life together in a population where none died before the upper limit of child bearing age and where the given set of fertility rates was in Operation".

SOCIO–ECONOMIC INDICES

Demographic Co-efficients : In analysing table of population data, indices which are designed to give a measure of population pressure in different regions are often found useful. Probably the simplest index of this type is obtained from the formula.

$$C = dR$$

Where C represents the demographic coefficient, pressure or demographic coefficient,

d the density of population.

R the net reproduction rate.

This expression is useful because it shows at a glance what future population densities might develop in different regions provided there were no migration, and where the current rates of mortality and fertility to remain unaltered. It helps to explain the building up of population pressure and migration trends, and is of use too in the study of economic problems.

Comparative Density : Population Pressure is not purely a function of density and replacement of population, because density of population is obtained by a consideration of the total area of the enumeration unit. This area may vary in its capacity to support population according to its geographical character. Hence more refined indices have been designed to eliminate this anomaly and to illustrate comparative density. The simplest index of this type is expressed by the formula.

$$D = \frac{P}{S}$$

Where D = Comparative density,
P = the total population,
S = the total cultivable land.

But cultivable land in itself varies, according to position, soil and other factors. To consider all the implications and to express them satisfactorily in the form of a simple numerical index is formidable undertaking, particularly as problems of optimum population remain as yet unsolved. However, it is possible to refine further the above index by weighting the figure for cultivable land.

Other Coefficients stress the relationship between the population of an area and all forms of income in that area. In this context regional variations can be

$$\text{Population Pressure} = \frac{\text{Total Population}}{\text{Total Income}}$$

conveniently mapped by relating local coefficients to the national coefficient as an index, i.e.

$$\text{Index} = \frac{\text{Total Local Population / Total National Population}}{\text{Total Local Income / Total National Income}} \times 100$$

Such maps are to use in the study of internal migration and regional development. In the field of social area analysis, a wide variety of indices and indicators are in use to measure such things as standard of living, social class, ethnic segregation and urbanization. Many of these have been employed by geographers.

The Crude Birth Rate

The crude birth rate is usually expressed as the number of children born annually per thousand of population, obtained from the formula :

$$\frac{\text{Number of birth}}{\text{Total Population}} \times 1000$$

The Crude Death Rate

The crude death rate is also expressed as an index—the number of deaths per thousand of the population, obtained from the formula :

$$\frac{\text{Number of Deaths}}{\text{Total Population}} \times 1000$$

This is not a very refined measure of mortality because the age and sex structure of the population is not taken into account in its calculation.

Fertility Ratio

It is difficult to compute the standardized rates, however, there is a more simple method of arriving at a good estimate of fertility by relating the number of young children in the population to the number of women of child-bearing age. This fertility rates is usually expressed as :

$$\frac{\text{Number of children under 5 years old}}{\text{Number of women aged 15-50}} \times 1000$$

The information necessary for the calculation is obtained directly from the tabular data recorded in various census reports.

QUANTIFICATION IN SETTLEMENT GEOGRAPHY

It becomes imperative to devise some means of quantitative analysis of the settlement pattern, and students of settlement have long concerned themselves with the problem of evolving formula for this purpose. In considering dispersion and concentration of settlements, for example, a number of variables have to be considered. These include :

(a) number of settlements
(b) number of houses in each settlement
(c) size of population in each settlement
(d) area of the region served by each settlement
(e) distance apart of settlements.

No one formula has ever been satisfactorily devised which would take all these variables into consideration, but a number of indices, in which two or three variables are involved, may be used for mapping purposes and certain distributional aspects.

1. Demongeons Co-efficient of Dispersion

This is obtained from formula :

$$C = \frac{E \times N}{T}$$

where E = Population of a place
N = Number of places
T = Total population.

2. Simple Index of Dispersion

Unfortunately, some of the censuses do not differentiate between chief place and isolated settlements in any one parish, but some idea of settlement dispersion may be deduced from the formula.

$$I = \frac{S}{H}$$

Where I = the index of dispersion
S = the number of settlements in any parish
H = the number of isolated houses.

A house which is situated a quarter of a mile from any other house is conveniently taken to be isolated.

3. Bernard's Index of Concentration

Bernard's formula which takes into account the number of houses (H), the area under consideration (A) and the number of settlements (S).

He argues that concentration of settlement is a function of the mean number of houses in each settlement *H/S* and of the number of settlements in a given area *A/S*. Considering both functions, he obtains the formula :

$$C = \frac{(H)\ (A)}{S\ \ S} = \frac{HA}{S^2}$$

4. Kant's Index of Concentration

A most indigenous formula to show diversity of settlements was devised by E. Kant to illustrate settlement. It was designed to be used to reduce a map showing distribution of habitations by means of non-quantitative of symbols in which dispersion and concentration of settlement was more precisely reflected in terms of distance between the habitations. The formula is :

$$X = \frac{I}{M} / \frac{A}{D}$$

Where *X* is the relative interval between two settlements,

I/M = the scale of the map.
A = the equal to the area under consideration.
D = the diversity of habitations.

5. Near Neighbourhood Statistics

This is a value arrived at statistically and is based on near neighbourhood analysis. It indicates the degree to which the settlements of any area are agglomerated or dispersed either uniformly or at random.

6. Settlement Groupings

The size of settlements and the grouping of houses in settlements may be indicated by adopting the following coefficient of grouping (*C*).

$$C = \frac{\text{Number of inhabitants}}{\text{Number of settlements}}$$

Population totals in this case are taken and the number of settlements have to be counted from contemporary topographical maps. Another coefficient (CI) is as follows :

$$CI = \frac{\text{Number of houses}}{\text{Number of settlements}}$$

Information about number of houses is made available from the census of some of the countries.

7. Density of Housing

The diversity of housing of any place is conventionally expressed as the ratio of the number of houses or alternatively the number of habitable rooms (i.e., rooms, kitchen, halls and bathrooms) per unit of area. But this ratio, however, is not always reliable, because it is very often invalidated by the inclusion of extensive areas of parkland taken up by industrial sites, countries and so on. For some purpose a comparative density of housing is useful, which is expressed by the ratio of the number of houses per unit of area of land used for residential purposes only.

8. Multivariate Analysis of Urban Areas

There are number of approaches to the study of typology of social areas than social area analysis. For instance, if matrices of correlation coefficients between each plain of all the variables selected for study are constructed it becomes apparent that causal relations between any single independent and a number of dependent variables is not indicated. The matrices in fact exhibit different degrees of correlation between the variables selected. In such circumstances component analysis provides a particularly useful approach to the study of a large universe of variables where the object is to collapse them into fewer and more manageable groups.

9. Statistical Approach to Town Analysis

A number of statistical approaches to the study of urban geography was recently started to analyse the problems of functions of towns, size of towns, distribution and hierarchy of towns. The data is largely available from field surveys as well as from census records. The aim has been to establish the nature of relationship between population size, functional diversity and spacing of towns. Such analysis have called for more precise definitions of terms, particularly of function, functional unit, establishment, central place, etc.

QUANTITATIVE METHODS IN ECONOMIC REGIONALIZATION

Quantitative methods have become an indispensable tool in the hands of Indian economists, planners and especially the geographers. The major focus and thrust field of geography is to analyse any phenomena in terms of regionalization in order to measure the existing variations both in space and time. The early surveys and analysis on the problem of regionalization was based on a simple description, the statistical approach is predominant in the present-day works. Resource combination regions for the establishment of any industry is important for a successful and economical location in any such regions. Without the statistical means it is not possible to analyse this problem in a more convincing manner.

The term 'region' is widely used in recent years with reference to a great variety of problems. It is an area homogeneous in respect of a particular set of conditions determined by the purpose for which it is delineated. For regional economic planning a region is a homogeneous unit with unique advantages of developing some kinds of production specialization. For such type of analysis quantified methods helps in a more meaningful manner.

The quantitative methods have found a wider application in economic geography since the establishment of a geographical section—Regional or Survey Unit—at the Indian Statistical Institute. After having thoroughly studied the various experiments in the field, M.N. Pal (1961)[6] had made an attempt to delineate the areas of different levels of economic development, determined with a complex set of indices which are recommended to be used in regional planning for balanced regional development. The basic difficulty in compiling these indices lies in the proper selection of their initial components and specific weights. Quantitative methods are extensively used in India especially to find out the distinctions between regional levels of development. A complex mathematical analysis of various initial statistical data used in the monograph entitled 'Levels of Regional Development in India' by Ashok Mitra.[7]

In the year 1966 Prof. Berry[8] applied some mathematical methods in economic regionalization of India. He used the multiple factor analysis while computing data on "Accounts Relating to the Inland Trade of India for twelve months ending March 1960". Prof. Berry explored the interrelationships between commodity flows and spatial economic structure. When the computed data are plotted on map they revealed India to be partitioned into four major 'functional economic regions' with the metropolitan centres of Calcutta, Bombay, Madras and Delhi (Commodity flows expressed by quantity) or unto three functional regions with the centres of Calcutta, Bombay, Madras (Commodity flows expressed by value). Although this study has a high value, yet it suffers from some defects both of general and particular nature.[9] In particular, the high degree of generalization, static nature of regionalization, exaggerated role conferred on the three main posts and others draw the attention of the reader. However, the significance of applications of mathematical methods in economic regionalization will undoubtedly grow very rapidly in future. The Soviet Geographers also studied the problems of economic regionalization especially in the case of India. The Institute of Geography, Academy of Sciences, USSR, under the general guidance of Prof. K.M. Popov has published a series of monographs called "Regions of India".[10]

In the year 1960, Galina Sdasyuk, a Soviet Geographer was given an opportunity by P.C. Mahalanobis, Regional Survey Unit, I.S.I. to explore the possibilities of applying Soviet experience to Indian conditions.[11] At the Indian Statistical Institute in 1961 Professor Komar, U.V. and K.M. Popov investigated the problems of economic regionalization of India. In the year 1962 the well known Soviet Geographers Prof. A.M. Ryabchikov and V.A. Nikolaev conducted a study at the Indian Statistical Institute on the problems of physical geographical regionalization of India and its Relationship with Economic Regionalization of the country. A joint study was also conducted by Bhat, L.S.; Komar, I.V.; Pal, M.N.; and K.M. Popov on the problems of application of Science and Technology for the benefit of less developed areas which is to be presented at United Nations Conference (Geneva).[12]

It is also very interesting that the Soviet and Indian scientists have common views on the problems of economic regionalization and regional planning. The methods and concepts introduced have largely been used for the planning purposes.

It is a major and main thrust of the field of geography is to regionalize any phenomena for planning purposes. The problems connected with eradication of economic backwardness and industrialization in India call for a most rational location of new enterprises and this is not possible without studying the nature of formation of economic regions. As a result, the problems of economic regionalization of India occupy a prominent place in Indian geography. Quantitative methods are to be used for the formation of economic regions.

REFERENCES

1. Miller, A.A. (1953), *The Skin of the Earth*, p.43, London.
2. Goldberg, G.M. (1963), The Derivation of Quantitative Surface that from gross Sources, *Surveying and Mapping*, vol. 23, pp. 537-48, Washington, D. C.
3. Bakker, J. P.; Strahler, A. N. (1956), Report on Quantitative Treatment of Slope Recession Problems, p. 30, Amsterdam.
4. International Geographical Union (1956), Amsterdam, A Premier Report Containing 155 pp. in which a large number of articles, maps and bibliographies on the subject of slope is available.
5. Wentworth, C.K. (1930), A simplified Method of Determining the Average slope of Land surfaces, *American Journal of Science*, series 5, vol. 20, New Haven.
6. Pal, M.N. (1961), Quantitative Delimitation of Regions, *Bombay Geographical Magazine*, No. 8-9.

7. Mitra, A. (1964), Levels of Regional Development in India, *Census of India*, 1961, Vol. 1., New Delhi.
8. Berry, B.J.L. (1966), *Essays on Commodity Flows and Spatial Structure of the Indian Economy*, The University of Chicago, USA.
9. Andrzei Wrobel's Comments on this point are very interesting. See Andrzei Wrobel. "Prospectives and Limitations of the application of Mathematical Methods in Economic Regionalization". Proceedings of the 4th General Meeting of the Commission on Methods of Economic Regionalization of the International Geographical Union, 1965, Prague.
10. These Monographs have already been published, Northern India, Central India by O.B. Oskolkova; Western India, "From Ancient Rajputana to Modern Rajasthan" by L.A. Knyazhinkaya, Kashmir by V.A. Pulyarkink; South India by A.L. Batalov, L.A. Knyazhinkaya, Sdasyuk, G.V.
11. Galina, Sdasyuk, Basic Principles of Economic Regionalization and Economic Regions in the USSR, *The Indian Geographer*, Vol. 6, Dec. 1961. The History of Economic Regionalization of the USSR; *The National Geographical Journal of India,* Vol. VIII, June 1962; Economic Regionalization in India, *Sankhya,* Series B, Vol. 24, Part I and II, 1962.
12. Bhat, L.S.; Pal, M.N.; Komar, I.V.; Popov, K.M. (1963), Problems of Regionalization for the Development of National Economy, *Sankhya*, Series B, Vol. 25, Part 1 & 2, Nov.

NOTES

Weaver, J.C. (1954), Crop-combination Regions in the Middle West, *Geographical Review,* Vol. XLIV, pp. 174-200

Siddiqi, M.F. (1967), Combinational Analysis—A Review of Methodology, *The Geographer*, Vol. XIV, p. 83

Doi, K., (1959), The Industrial Structure of Japanese perfecture, Proceeding IGU Regional Conference in Japan, pp. 310-316

Rafiullah, S.M. (1965), A New Approach to Functional Classification of Towns, *The Geographer*, Vol. XII, pp. 40-53.

4

Sampling Techniques

SIGNIFICANCE IN RESEARCH AND DATA COLLECTION

Various numerical and mechanical techniques have been evolved to solve the problems of Vast Wealth of statistical data. Even the techniques are also evolved to make the observations representative by using the small proportion of the available statistics in the form of case studies as the reliable unit. The statistical sampling is the best way to find out the more reliable picture instead of the generalization. On the other hand, selective studies which do permit certain generalizations may be undertaken if a statistical sampling basis is adopted for the work. In fact, this is in many cases preferable to a full survey of a particular problem, for reasons available time, manpower and finance.

Three major methods have been evolved to carry out the sample surveys. The most easiest one so far is the random sample. Any object can be studied either it is farm or village or any unit. For a geographical point of view the random sample may be inappropriate for the investigation of spatial variations of some phenomena, since it can lead to a biased sample if there is marked spatial clustering of particular categories of the features which are being studied. For example, the landscape elements are rarely scattered at random and consequently one must be cautious enough in exercising the use of these methods. Therefore, it is advisable to use the stratified random sample technique. The unit to be studied must be selected from the strata formed by any phenomena such as the study of settlements and occupational structure can be stratified and on random basis the unit of study may be selected. But the method of stratified samples requires considerable prior information about the object of the study.

The method of systematic sampling is taken at regular intervals by dividing the area into equal units. It needs no prior information like stratified random sampling. If a farm is selected from the large size of the farming areal units, therefore, a farm to be selected should also be large.

In Indian context most of the socio-economic problems are so large in spatial term that it is impossible to cover all the units of a variable. Secondly, socio-political and economic constraints are so rigid alone with the limitations of resources and trained personnel that there is no option except to opt for the sampling techniques of various types for a better planning well in time.

The basic logic behind various types of sampling is to find out the representative unit for the region as a whole so that the results obtained can be related with the results obtained at various micro-meso and macro level. The different methods of sampling which ensure better representation of any problem under different situations are as follows :

Random Sampling

The type of sample is most simple and every unit can be selected for the purpose. It is most useful and successful in a region where homogeneity of any unit prevails either it may be the size of the population or village or town etc. In this sampling every member of the universe (Population is supposed to be selected but this method is useful only when there exists homogeneity with each member of the universe. The simple procedure is well known as it is just like taking out a unit sample in a lottery manner.

Systematic Sampling

It may be unjust to cover and study the area through random sampling because many times a random sample does not have the good coverage of the area of study. The basic concept of the systematic sampling is to select the sample at regular interval. For example if ten farming fields are selected, should get the chance to

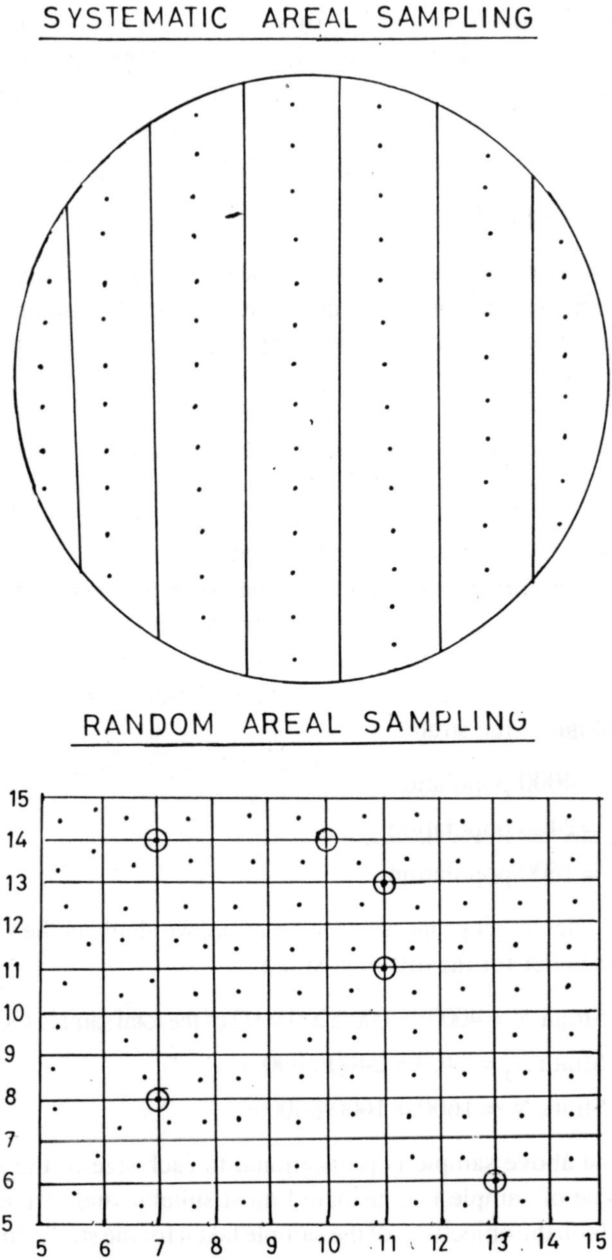

FIG. 4 : Sampling Techniques

be selected at regular interval and distance such as every fifth fields if the total number of fields are fifty because ten farming fields are going to be selected. Sometimes it is criticized that the systematic sampling is more or less is a random sampling. Systematic sampling has certain plus points over the random sampling because it is evenly distributed over the space and universe.

Stratified Sampling

In order to find a representative sample stratified sampling is used. The basic condition of this sampling is to find out the homogeneous strata and from each strata desired number of units are selected with the care that selected unit is more and more representative of the strata. The strata should be formed on the basis of the common characteristics of the items to be put in each stratum. This means that various strata be formed in such a way as to ensure elements being most homogeneous within each stratum and most heterogeneous between the different strata. Thus, strata are purposively formed and are usually based on past experience and personal judgement of the researcher. A very sound analysis is needed for the information of strata. For example if we want a sample size of 30 out of the size of population of 8000 which is divided into three strata :

$N_1 = 4000$ population

$N_2 = 2400$ population

$N_3 = 1600$ population

In order to get proportional allocation, we shall get the sample sizes as under for the different strata :

1. Strata $N_1 = 4000 = 4000/8000 = 1/2$ of the total sample (30) = 15
2. Strata $N_2 = 2400 = 2400/8000 =$ " " " " = 9
3. Strata $N_3 = 1600 = 1600/8000 =$ " " " " = 6

(The above sample is proportional to each size of the strata. This type of sample is considered most suitable and efficient. It depends on the objectives of the sample taken for the study. In some cases equal number of sample selected from the different size of the strata is considered more efficient and meaningful).

Exercise

A population is divided into three strata so that $N_1 = 5000$ $N_2 = 2000$ and $N_3 = 3000$. Respective standard deviations are :

1 = 15, 2 = 18, 3 = 5

How should a sample of size n = 84 be allocated to the three strata, if we want optimum allocation using disproportionate sampling design ?

Solution

Using the disproportionate sampling design for optimum allocation, the sample sizes for different strata will be determined as under :

Sample size for strata with $N_1 = 5000$

$$n_1 = \frac{84\ (5000)\ (15)}{(5000)\ (15) + (2000)\ (18) + (3000)\ (5)}$$

$$= \frac{6300000}{126000} = 50$$

Sample size for strata with $N_2 = 2000$

$$n_2 = \frac{84\ (2000)\ (18)}{(5000)\ (15) + (2000)\ (18) + (3000)\ (5)}$$

$$= \frac{3024000}{126000} = 24$$

Sample size for strata with $N_3 = 3000$

$$n_3 = \frac{84\ (3000)\ (5)}{(5000)\ (15) + (2000)\ (18) + (3000)\ (5)}$$

$$= \frac{1260000}{126000} = 10$$

Cluster Sampling

A cluster sampling is made from a very large area as the area is divided into smaller areas. From each smaller area sample units are selected from each small areas located adjoining to large number of smaller areas. Cluster sampling is useful because it requires less cost of surveys. But it is less accurate as compared to the random sampling. For instance if one is interested to find out the agricultural performance of Madhya Pradesh districts where most of the districts performs uniformly. So it is better to select few adjoining districts from each direction of the State.

Purposive Sampling

In purposive sampling, the samples are selected to represent the universe. After deciding the general breakdown of the sample on scientific lines, the individual is at liberty to choose the actual sample unit. Purposive sampling is, therefore, a method of stratified sampling in which the selection within the stratum is non-random. Although, theoretically it seems less sound as compared to random sampling, but, with adequate safeguards, it can be made highly reliable and the extra cost of random sampling can be avoided.

It may, therefore, be said that the systematic purposive cluster sampling is most suitable for the purposes of landuse surveys, human economic activities, rural development etc. For the sake of convenience and also in view of social, cultural and historical factors varying from one region to another, it would be better to undertake the surveys at regional level. After a detailed study of the factors affecting the major objective the region should be divided into various strata based on some homogeneous aspects and finally some representative villages may be selected where cluster of fields can be studied.

Exercise

A random sample of 100 farms from a region gives the average farm size of 50 acres with a standard deviation(s) of 26 acres. Can the average farm size in the region be taken as 60 acres?

Given $\overline{x} = 50$, $u = 60$, $n = 100$, we do not know the value of the population standard deviation. Since the size of the sample is large, the sample standard deviation(s) may be taken a good estimate of 0

Thus :

$$= [50 - 60] \quad = 10 \text{ acres}$$

$$3 \; S.E. = \frac{3 \times 26}{\sqrt{100}} = \frac{3 \times 26}{10} = 7.8$$

As the difference of sample mean and the given population mean (10 acres) is more than three times of the S. E. of mean (7.8 acres), the difference is significant at 1% level of significance. Hence, we cannot take the average farm size in the region as 60 acres.

The size of households in a locality is normally distributed. A sample of 10 households is drawn randomly and the number of persons in each household is found to be 14, 9, 8, 4, 14, 5, 11, 10, 12 and 3. In the light of these data can we consider the mean size of households in the locality as 12 persons ?

Household	size (x)	$(x - \overline{x})$	$(x - \overline{x})^2$
1.	14	+5	25
2.	9	0	0
3.	8	-1	1
4.	4	-5	25
5.	14	+5	25
6.	5	-4	16
7.	11	+2	4
8.	10	+1	1
9.	12	+3	9
10.	3	-6	36
			142

$$x = 90 \quad (x - \bar{x})^2 = 142$$

$$\bar{x} = \frac{90}{10} = 9$$

$$S = \sqrt{(x-\bar{x})^2} = \sqrt{\frac{14.2}{10}} = \sqrt{142} = 3.768$$

Using the sencond expression for t (since only sample S.D. is known)

$$t = \frac{9-12}{3.768 / \sqrt{10-1}} = \frac{-3}{3.768 / \sqrt{9}} = \frac{-3}{3.768 / 3}$$

$$= \frac{-3 \times 3}{3.768} = \frac{-9}{3.9}$$

$= -2.3085$ taking absolute value $t = 2.3085$

The value of t for 9 d.f. at 1 per cent level of significance is 3.25 and at 5 per cent level of significance it is 2.26.

Thus the null hypothesis that the mean size of the locality is 12 persons cannot be rejected only at 1 per cent level of significance, however it is rejected only at 5 per cent level of significance. In other words we may conclude that the mean size of households in the locality cannot be 12 persons only if we agree to sacrifice some accuracy (i.e., hypothesis is rejected at 5 per cent level of significance), but more accurate conclusion would be that it might be equal to 12 persons.

NOTES

A detailed discussion of the applicability of several methods of stratification for agricultural output is given by W. B. Taylor and D. V. P. Clement. The New Zealand Agricultural Sample Survey, *Journal of the Royal Statistical Society*, pp. 409-424, London, 1956.

Monkhouse, F. J. and Wilkinson, H.R., (1971), Maps and Diagrams, Mithuen & Co. Ltd., pp. 472-73. An abundance of researches have been produced in developing these techniques such as Gregory (Statistical Methods and Geographers, London 1963) ; J. P. Cole and C.A.M. King (Quantitative Geography, London, 1968); J.P. Cole (Mathematics and Geography, *Geography,* Vol. 54, pp. 152-64, London, 1968) ; Poster, P.W. (A Bibliography of Statistical Mapping, Minneapolis 1964); Garrison, W.L. and D.F. Marable, (Quantitative Geography, New York, 1968); Yeats M. H. (An Introduction to quantitative) Analysis in Economic Geography, New York,1968); Berry, B.J.L. and Marble, D.F., (Spatial Analysis : A Reader in Statistical Geography, 1968 King L.J. (Statistical Analysis in Geography, 1969).

5

Measures of Inequality and Relationships

MEASUREMENT OF INEQUALITY

To measure the concentration, distribution and dispersal of any phenomenon different methods and techniques are used such as Location Quotient, Index of dissimilarity, Lorenz curve and Gini's Coefficient etc.

Location Quotient

In order to find out the proportion of any aspect in a district in relation to the state or country as a whole. We use the technique of Location Quotient. It is important to find out the relation of that aspect in terms of proportion may be important at local level and not at regional or country or state level. These proportion may be the percentage share of agricultural labourers to the total workers, share or proportion of scheduled castes to the total population, manufacturing workers to the total workers, proportion of population engaged in agriculture, industries and tertiary sectors etc. Location quotient which gives us the relative picture of such proportions is defined as the ratio of the proportion of a particular characteristic in an area to the same proportion in the region, symbolically it can be defined as :

$$LQ_i = \frac{P_{ij} / P_i}{P_j / P}$$

Where P_{ij} = number of persons in *j-th* (= 1, 2...*m*) category of area
$i = (1, 2 \ldots n)$

$$P_i = \sum_{i=1}^{m} P_{ij}$$ = total population in all the categories of area *i*.

$$P_j = \sum_{i=1}^{m} P_{ij}$$ = sum of persons of category *j* in all the *n* areas

i.e., population of region under category *j*

$$P = \sum_{i=1}^{n} \sum_{j=1}^{m} P_{ij}$$ = sum of P_i in all the areas i.e., total population

of the region in each category.

The analysis of location quotient is useful in regional analysis. In a regional context a higher or lower value of the location quotient indicates relative concentration or dispersion of the concerned attribute. This information is of limited use in exploratory work, but can be of considerable use in conjunction with other tools and techniques of analysis which fully recognise and incorporate in their framework of the broader prospects of regional analysis.[1]

Exercise

Find out the districts of relative concentration of agricultural labourers of the total workers.

Concentration of Agricultural Labourers

Districts	% Agricultural Labourers to the total workers	Location Quotient
1. Ajmer	8.32	0.89
2. Alwar	8.90	0.95
3. Banswara	7.35	0.78
4. Barmer	6.74	0.72
5. Bharatpur	8.97	0.96
6. Bhilwara	6.58	0.70

Districts	% Agricultural Labourers to the total workers	Location Quotient
7. Bikaner	6.77	0.72
8. Bundi	11.62	1.24
9. Chittaurgarh	0.76	0.08
10. Churu	4.17	0.44
11. Durgapur	7.42	0.79
12. Ganganagar	17.32	1.85
13. Jaipur	5.97	0.64
14. Jaisalmer	8.52	0.91
15. Jalore	14.69	1.57
16. Jhalwar	16.92	1.81
17. Jhunjhunu	6.60	0.70
18. Jodhpur	6.68	0.71
19. Kota	15.36	1.64
20. Nagaur	7.58	0.81
21. Pali	20.03	2.15
22. Sawai Madhopur	8.57	0.92
23. Sikar	5.31	0.57
24. Sirohi	20.63	2.21
25. Tonk	6.47	0.85
26. Udaipur	5.31	0.69
	9.31	1.00

Lorenz Curve[2]

It is used to measure the inequality in the distribution of wealth or income or any phenomena where inequality exists and needed to be studied. It deals with the cumulative percentage distribution of the two attributes at different points. The cumulative percentages of one variable upto certain points are plotted on a graph against the cumulative percentage of the other variable upto the same points. The different points obtained are then joined by a smooth freehand curve. For comparison a diagonal line is also drawn joining the last point and the origin, showing the line of equal distributions. The deviations of any curve from this diagonal is proportion to the level

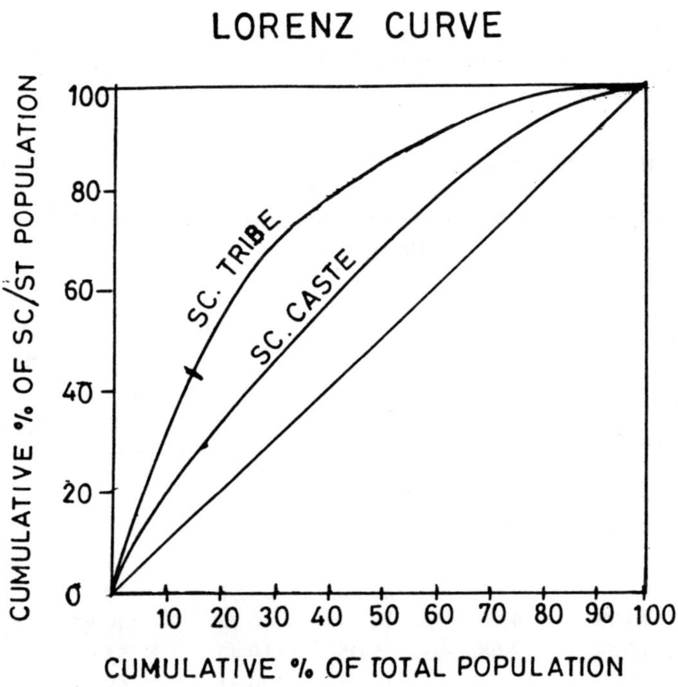

FIG. 5 : Lorenz curve

of inequality in the distribution of one attribute in relation to the other.

Firstly the data is arranged at descending or ascending order. Take the percentage of the total population of each state to the total population of the country. Note the percentage of the tribal population of each state to the total tribal population of the country.

The differences in the two concentration curves may be explained in terms of the underlying socio-economic, geographical and temporal processes.

Distribution of Scheduled Castes and Scheduled Tribes Population

Unit	%of ST Pop. of total Pop.	%distribution of total Pop.	%distribution of Tribal Pop.	Cumulative % distribution of total Pop.	Cumulative % distribution of Tribal Pop.
1.	88.76	0.09	1.24	6.09	1.24
2.	31.13	0.19	0.90	0.28	2.14
3.	28.98	0.27	1.22	0.55	3.36
4.	23.11	3.94	13.70	4.49	17.06
5.	20.13	7.47	22.66	11.96	39.72
6.	13.99	4.78	10.09	16.74	49.81
7.	12.84	2.68	5.05	19.42	55.00
8.	8.75	10.09	13.33	29.51	68.33
9.	7.71	1.14	1.32	30.65	69.65
10.	7.93	1.98	2.20	32.63	71.85
11.	5.86	9.03	7.98	41.66	79.83
12.	5.72	7.94	6.84	49.60	86.67
13.	5.15	7.41	5.74	57.01	92.41
14.	4.10	0.62	0.38	57.63	92.79
15.	3.81	7.80	4.48	65.43	97.27
16.	1.26	3.82	0.73	69.25	98.00
17.	1.21	4.62	0.84	73.87	98.84
18.	0.79	5.25	0.62	79.12	99.46
19.	0.23	15.83	0.54	94.95	100.00
20.	0.00	1.80	-	96.75	100.00
21.	0.00	0.80	-	97.57	100.00
22.	0.00	2.43	-	100.00	100.00

Correlation

The relationship and association between two variables either independent with dependent is known as correlation. The degree of relationship has been measured with the help of quantitative data as well as with the graphic representation. The reliability of quantitative measurement of this exercise (correlation) is widely accepted as compared to the graphic representation. For the purposes of quantitative measurement various techiques have been evolved.

The identification of the causal relationships among the different characteristics of any study is an essential concern of a scientific investigation. A causal relationship between the two characteristics exists only when one of them may logically be considered as the cause of the other. The factor which is supposed to be the cause is known as the independent variable and the one which is supposed to be the effect is known as the dependent variable. Thus the variations in the independent variable may be explained in terms of the variations in the independent variables. In case there exists a causal relationship, both the values of independent and dependent variables will vary together. This property of co-variation is also termed as correlation.

If an increase in the independent variable also tends to cause an increase in the dependent variable, the correlation is said to be positive. For instance, in a water scarcity area such as Rajasthan, a higher amount of rainfall may tend to cause a higher amount of crop production also, or better facilities of irrigation may lead to higher consumption of fertilizers. On the other hand if an increase in the independent variable tends to cause a decrease in the values of the dependent variable, the correlation is said to be negative. Another example of negative correlation could be between higher the medical facilities and higher the infant mortality rate. In case, there is no logical basis for a correlation between variables, it should not be taken as a causal relationship. Any such relationship is faulty and should not be considered for the study. The interference of unknown variable may also cause such faulty relationship.

The measurement of the degree and direction of correlation helps particularly the geographers in explaining the variations in

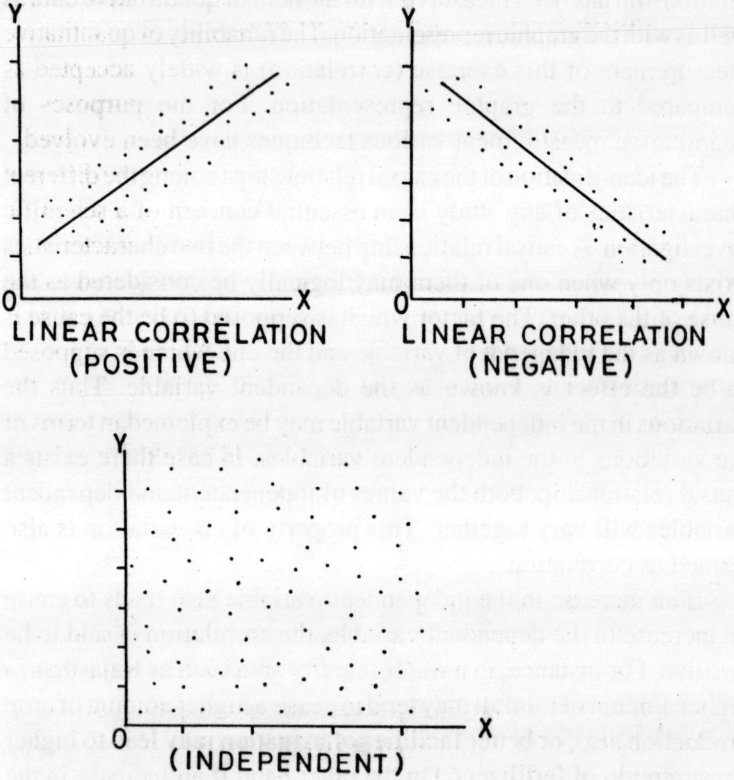

FIG. 6 : Relationship between Dependent (Y) and Independent (X) Variable

various spatial phenomena. To represent the correlation between two variables visually. We consider the values of the independent and the dependent variables as X and Y coordinates of a two dimensional space. Each pair of observations can then be plotted on the graph paper, according to its X and Y coordinates. The graph so obtained is called as scatter diagram. Such a diagram is useful in giving us an intuitive idea of the relationship between independent variable X and the dependent variable Y (Fig.6). If the values of X and Y vary such that each point falls exactly on a straight line, the relationship is said to be linear. The higher the deviation of the points (dots) from the straight line the weaker will be the correlation between X and Y variables. Similarly the positive and negative correlation depends on the increase and decrease of the dependent and independent variables.

METHOD TO MEASURE THE LINEAR CORRELATION

It has been mentioned earlier that the method to measure the relationship through the scattered diagram gives only the general picture but a quantitative picture will be more reliable and acceptable.

A precise quantitative measurement of the degree and direction of a linear correlation was propounded by Karl Pearson such as :

$$r = \frac{XY - \dfrac{X \times Y}{N}}{\sqrt{X^2 - \dfrac{(X)^2}{N}} \sqrt{Y^2 - \dfrac{(Y)^2}{N}}}$$

This measurement is known as the product moment correlation coefficient or simple correlation coefficient. By symmetry it is clear that $r\,xy = r\,yx = r$.

General Characteristics of Correlation Coefficient

(A) If the sign r is positive the variables x and y are positively related and if the sign is negative, they are negatively correlated.

(B) The value of r varies between -1 and $+1$. The value $+1$ or -1 indicates a perfect positive or negative correlation. As the extent of correlation decreases the value of r approaches zero.

Test of Significance of Correlation Coefficient

Based on a smaller number of observations a coefficient of correlation is considered a simple correlation. Using the test of significance of *r*, it is possible to infer whether the correlation coefficient of the bivariate normal population (the correlation between the same variables but based on a fairly large number of observations) will be zero or not. Under the null hypothesis, that the population correlation is zero, the expression as given below will be followed by '*t*' distribution with (*n*–2) degrees of freedom.

$$t = r\sqrt{\frac{n-2}{1-r^2}}$$

Where '*n*' is the number of observations used and '*r*' is the coefficient of correlation.

If any computed value of '*t*' is less than the corresponding tabulated value the correlation coefficient is said to be insignificant, meaning thereby that over a larger number of similar observations the two variables will be independent, as the hypothesis that the population correlation coefficient is zero is accepted. Whatever the simple correlation in this case we get may be attributed to the chance factors only. On the other hand if the computed value of '*t*' is greater than the tabulated '*t*' the correlation co-efficient is said to be significant and the population correlation coefficient between the two variables in this case is not considered to be zero. It is obvious from this test that the significance of the correlation coefficient is directly proportional to not only *r*, but also to (*n*–2). In some cases, because of a larger value of '*n*' a smaller correlation co-efficient may become significant, whereas in some other cases, a larger correlation coefficient may become insignificant because of the smaller value of '*n*'.

Exercise

The percentage of Scheduled Castes population to total population and the percentage of landless agricultural workers are given below for 12 randomly selected villages of a district. Does there exist a linear correlation between them in the district.

Village Code	% of Scheduled Castes of total population (X)	% of landless labourers to total agricultural workers (Y)
1.	2.8	38.9
2.	8.6	33.7
3.	1.8	40.6
4.	4.8	22.0
5.	4.8	29.5
6.	2.0	4.9
7.	6.7	37.6
8.	2.2	2.6
9.	5.7	28.2
10.	7.3	33.2
11.	6.3	65.8
12.	2.6	3.7

Solution through Karl Pearson formula mentioned above

Village Code	X	Y	X^2	Y^2	YX
1.	2.8	38.9	7.84	1513.21	108.92
2.	8.6	33.7	73.96	1135.69	289.82
3.	1.8	40.6	3.24	1648.36	73.08
4.	4.8	22.0	23.04	484.00	105.60
5.	4.8	29.5	23.04	970.25	141.60
6.	2.0	4.9	4.00	24.01	9.80
7.	6.7	37.6	14.89	1413.76	251.92
8.	2.2	2.6	4.84	6.76	5.72
9.	5.7	28.2	32.49	795.24	160.74
10.	7.3	33.2	53.29	1102.25	242.36
11.	6.3	65.8	39.69	4329.64	414.54
12.	2.6	3.7	6.76	1437.41	98.54
Total	55.60	374.09	317.08	14759.58	1902.64

$$r = \frac{XY - \dfrac{X \times Y}{N}}{\sqrt{X^2 - \dfrac{(X)^2}{N}}\sqrt{Y^2 - \dfrac{(Y)^2}{N}}}$$

$$1902.64 = \frac{\dfrac{55.60 \times 374.90}{12}}{\sqrt{317.08 - \dfrac{55.6 \times 55.6}{12}}\sqrt{14759.58 - \dfrac{374..9 \times 374.9}{12}}}$$

$$= \frac{165.6}{\sqrt{59.47}\sqrt{3047.8}}$$

$$= \frac{165.6}{7.71 \times 55.20}$$

$$= \frac{165.6}{425.59} = +0.389$$

On the basis of the above mentioned exercise it is evident that there exists a positive relationship between the percentage of scheduled castes to total population and percentage of landless labourers to total agricultural labourers in the present sample of 12 villages. However, without applying the test of significance, we cannot generalize this positive relationship for all the villages in the district as the number of observations in the present case is very small. The test if carried out in the following manner.

$$t = r\sqrt{\frac{n-2}{1-r^2}} = .389\sqrt{\frac{12-2}{1-(.389)^2}}$$

$$= .389\sqrt{\frac{10}{.8487}}$$

$$= .389\sqrt{11.784} = .389 \times 3.4324 = 1.3353$$

Here we may conclude that even there exists a positive correlation in the 12 sample villages, but it is small in relation to the size of the sample and so may be attributed to the chance or random factors. Thus on the basis of this correlation co-efficient we cannot conclude that the data supports a causal relationship between the two variables in whole of the district.

Rank Correlation

For a rank correlation data may be collected either in ranked form or actual quantitative values may be converted into the ranked form. The index of rank correlation *rh o* (*p*) is given as below :

$$rho\ (p) = 1 - \frac{6\ di^2}{n^3 - n}$$

Where '*n*' is the total number of observations, and *di* is the difference in the ranks of the two variables in the *ith* observation.

Exercise

In the exercise given above if the variables *X* and *Y* are converted into their ranks. *Rx* and *Ry* respectively as given below. The co-efficient of rank correlation may be computed in the following manner :

Village Code	*Rx*	*Ry*	*di – Rx – Ry*	*di²*
1.	8	3	5	25
2.	1	6	-5	25
3.	12	2	10	100
4.	6.5	10	-3.5	-12.25
5.	6.5	8	-1.5	2.25
6.	11	0	0	0
7.	3	5	-2	4
8.	10	12	-2	4
9.	5	9	-4	16
10.	2	7	-5	25
11.	4	1	3	9
12.	9	4	5	25
				247.50

$$= 1 - \frac{6 \times 247.5}{12^3 - 12}$$

$$= 1 - \frac{1485}{1728 - 12}$$

$$= 1 - \frac{1485}{1716}$$

$$= 1 - .8653$$

$$= 0.1347$$

REGRESSION ANALYSIS

The problem of regression is discussed when the relationship between two or more variables are established. While degree of relationship is established between two or more variables is discussed under correlation. In historical perspective the concept of regression was first discussed by Galton Francis during the period 1822 to 1911, after this period the regression technique became too common for a variety of problems and a number of regression was established and named such as linear or curvilinear. Linear relationship between two variables is represented by a straight line which is known as regression line. The line of average relationship is another name for a regression line. In the study of linear relationship between two variables *Y* and *X*, suppose the variable *Y* is such that it depends on *X*, then we call it the regression line of *Y* on *X*. If *X* depends on *Y*, it is called the regression of *X* and *Y*.

There exist a relationship between a dependent variable Y and an independent variable *X*, the points of the scatter diagram will move around a curve which may be called the curve of regression. But when the curve is a straight line it is said to be the line of regression. Through a scatter diagram one can draw a large number of straight lines. In these equations of straight lines, if we put different given values of *X* we get the corresponding estimated values for *Y*. These estimated values of *Y* would be different from the actual values of *Y*. Out of all these lines the line which gives the minimum sum of squares of differences between actual values of Y and their estimated values is supposed to be the best suitable line and is known as the line of regression of *Y* on *X*. The method by which we determine such equation is known as the least square method.

The basic objective of a regression analysis is to estimate the values of *'a'* and *'b'*. The estimated least square form of the relationship is given by Johnson[3] in 1972 as following :

$$Y = a+bx$$

Where $b = \dfrac{XY - \dfrac{XY}{N}}{X^2 - \dfrac{(X)^2}{n}}$ is an estimate of *b*.

and $a = a = \overline{Y} - b\overline{X}$ is estimate of *a*.

The scatter diagram in relation to the regression line, helps us is judging whether the assumptions are justified or not. Once it is established that the assumptions are not seriously violated one can use the line of regression for further analytical purposes.

Exercise

Find out the functional form of linear relationship between rainfall and agricultural productivity. What is the average rate of change in agricultural productivity with respect to rainfall and how good is the linear relationship.

Unit	Y	X	Y^2	X^2	XY
A	364.6	101.3	132933.16	10261.69	36933.98
B	402.1	86.9	161684.41	7551.61	34942.49
C	416.1	99.7	173139.21	9940.09	41485.17
D	310.6	87.9	96472.36	7726.41	27301.74
E	536.2	94.4	287510.44	8911.36	50617.28
F	367.5	64.3	135055.25	4134.49	23630.25
G	448.7	85.3	201331.69	7276.09	28274.11
H	420.7	89.2	176820.25	7596.64	37508.60
I	287.7	70.2	82771.29	4928.04	20196.54
J	366.1	52.1	134029.21	2714.41	19073.81
K	596.0	135.4	355216.00	18333.16	80698.40
	4516.1	966.7	1936964.27	89733.99	410662.37

$$b = \frac{410662.37-(4516.1 \times 966.7)/11}{89733.99 - (966.7)^2/11} = \frac{13779.29}{4778.64} = 2.88$$

$$a = \frac{4516.1}{11} - 2.88\,\frac{(966.7)}{11} = 410.66 - 253.09 = 157.46$$

Thus the regression equation of Y on X is

$Y = 157.46 + 2.88\,X$

The slope of the regression line b which is also known as regression co-efficient, shows the estimated average change in Y with respect to X. Thus from the regression equation we can say that the relationship between agricultural productivity and rainfall is such that a centimetre change in rainfall by and large causes an increase of 2.88 in the agricultural productivity. The result indicates that a great deal of agricultural productivity depends on rainfall.

REFERENCES

1. Florence, P.S., Investment, *Location and Size of Plants*, Cambridge University Press, 1948. Also for a detailed analysis see Isard, W. *Methods of Regional Analysis*, M.T. Press, pp. 123-126.
2. Lorenz M.O., 'Method of Measuring the Concentration of Wealth, Quarterly Publication of the *Am. St. Ass*, p. 209-19, 1905.
3. Johnson, J., *Econometric Methods*, MacGraw Hill, 1972, pp. 8-16.

6

Sources of Statistics and Statistical Organizations

PRE - AND POST-INDEPENDENCE PERIOD

Statistical information about a place or region, state or nation is necessary for running the administration efficiently. Its origin dates back to the period of *Ramayana* when statistics regarding employment and taxes, were collected during the rule of Raja Bharat. Kautilya's *Arthashastra* also throws light on the collection machinery of statistical information regarding land, area, agricultural production, population, taxes etc. During the Moghul period, statistics were collected and used for administrative purposes. The main documents related to statistical information are *Tuzke-Babri* and *Ain-i-Akbari. Ain-i-Akbari* is considered to be one of the very informative and standard documents of Emperor Akbar's period. From the details in *Ain-i-Akbari,* it is evident that there has been an organised machinery for collecting statistical information.

After the Moghul period, the British period started in India. *Ryotwari* system was introduced in some parts of India during the 18th century and the land revenue officers were appointed by East India Company to collect land revenue. The revenue rate was based on agricultural production and fertility of soils. The procedure continued till the existence of Zamindari System in India.

The need for statistical information was felt more and more in the nineteenth century due to the recurrence of famine in 1860 which is a historical famine of India. Till that time, government had no knowledge about the population and its food requirements. A partial census in India took place in 1872. The first census covering the whole of India, was conducted in 1881. Since then, decennial census is a regular feature. The famine commission of 1880 recommended the creation of the department of agriculture and some posts of statistical officers. The statistics of foreign trade, prices and industries, on the other hand, were collected and published by the Department of Finance and Commerce. This led to the publication of "Agricultural Statistics of British India" in 1886.

In order to collect and compile enormous data at the state level, a Statistical Bureau was established at the centre in 1895. Besides agricultural industries. In 1905, the Directorate of Commercial Intelligence and statistics was established and statistical Bureau was merged with it.

Immediately after the report of the Economic Enquiry Committee of 1925, the report of the Royal Commission on Agriculture was out in 1928. The Royal Commission recommended the constitution of an Imperial Council of Agricultural Research, to coordinate agricultural research and collect data about agriculture and animal husbandry. An important event that occurred in 1933 was the establishment of Statistical Research Bureau. Another committee was set up in 1934 known as Bowley Roverston Committee. The committee recommended and gave details of the census of the production and the measurement of the National Income. It also recommended the appointment of a Directorate of Statistics to collect data. In 1939, there was an outbreak of World War II. The need for statistical information was felt more and more. The Department of Industries and Civil Supplies meant for controlling the civil supplies was entrusted with the job of collection of the statistical information, relating essential commodities to ensure completeness of data on industries, an act known as Industrial Statistics Act was passed in 1942. This act compelled the industries to furnish the required information for statistical purposes. More and more developments took place from time to time, regarding the collection of statistical information, making the data more complete and reliable. Also an effort was made to collect statistical data, in varied fields, related to economy and manpower. Since statistical information has been the backbone of government activities and planning, it has been paid great attention after the independence of India in 1947.

Indian government has a federal set-up as far as the administration of the country is concerned. Hence, there is a combined responsibility of the states, as well as the centre for internal matters. In view of this, all the jobs are taken either by state government or by the central government or by both. Different categories of works are covered by different ministries of the states or the centre. For a systematic organization and the collection of statistical information, Government of India established different departments at the centre and state levels. The Cabinet secretariat has a Department of Statistics. One wing of this department is National Sample Survey Organization (NSSO) set up in May, 1951. These two are important statistical wings of the Government of India and hence it becomes necessary to know about them in details.

CENTRAL STATISTICAL ORGANIZATION (CSO)

Central government established central statistical organisations, under cabinet secretariat, with the objective of creating coordination of large variety of statistical information, collected at the centre and state level. It performs many more functions as listed below :

1. Coordination of statistical activities at the centre and the state.
2. Advisory work concerning the statistical matter, particularly standardization of concepts and definitions to maintain uniformity, throughout the country.
3. Collection of statistical data related to planning.
4. Training of statistical personnel.
5. Compilation of national income estimates.
6. To provide statistical data of the nation to the United Nations statistical offices and other international institutions.
7. To plan and coordinate the conduct to the annual survey of industries and publish the results.
8. To attend to the work of International Statistical Institutes (conferences) held in India and abroad.
9. The display of charts and graphs pertaining to the national data, which are of administrative interest.
10. Circulation of regular publications.

This organisation does many other things concerning statistical matter, needed from time to time. Each of the above mentioned functions, is looked after by a separate division with its head-quarters in Delhi, except the Industrial Statistics Wing at Calcutta. We give here the brief account of the above mentioned functions.

The CSO examines various schemes running at the state level and makes specific suggestions for their improvement. Now CSO has a network all over the country. CSO coordinates statistics collected by different ministries. It also coordinates with National Sample Survey Organisation.

The Central Statistical Organization, supplies statistical information about India to United Nations Statistical Office and International Agencies. It provides data for publication in (i) The U.N. Monthly Statistics, (ii) U.N. quarterly Bulletin on Commodity Trade Statistics, (iii) The U.N. Demographic Year Book, (iv) The Economic Council for Asia and Far East (ECAFE), Quarterly Bulletin and Annual Surveys. Besides these, CSO supplies compiled data of many other international agencies for publication. It also supplies information regarding the methodology adopted in the collection of data, the coverage and the scope.

The Directorate of Industrial Statistics was set up in 1944 at Calcutta. But later in 1957, it was merged with Central Statistical Organization. Its main function has been the annual census of manufacturing industries, which is known as the Annual Survey of Industries (ASI). This unit of CSO has various functions such as (i) to coordinate the conduct of annual survey of industries in all the states, and process the data for publication, (ii) to giving monthly account of the production of selected industries through a monthly bulletin, (iii) to compute and publish the monthly index of industrial production, obtained for selected industries and (iv) to function in many other ways. The CSO renders advice to various ministries, in the central government. It examines the bulletins of the state level. It lays down the definitions and concepts in conformity, with international standards e.g. International Standard Industrial Classification (ISIC) ; International Standard Classification of Occupation (ISCO) and Standard Occupational Classification (SOC) for India. The adoption of standard definitions and classification makes the information comparable within the country and at the international level.

A planning cell was set up in CSO in 1955, to maintain liaison between the Indian Statistical Institute and the planning division. All statistical matters related to planning are dealt within this wing.

The CSO conducts regular training courses at its offices to train the personnel in statistical services throughout the country.

The CSO makes national income estimates for the centre as well as the states, in collaboration with various statistical agencies at the state level. A National Income Unit (NIU) was created in 1955. The unit exclusively collects the data, concerning the national income. A white paper on national income is published by an industrial intelligence unit, with regard to industries in India. This unit assists the private and public sectors in many ways, particularly, giving information about imports, exports, demand etc.

The pertinent information of administrative and public interest is displayed through charts, graphs and diagrams. Such diagrams are displayed in an exhibition hall maintained by the CSO.

As a matter of fact, statistical information is collected and compiled in a vast field of economic activity like motor vehicles, civil aviation, foreign trade, inland trade, taxation and revenue agriculture forest fishery, livestock and what not. All this compiled information is published in one or another form in the journals and bulletins. We enlist few publications of CSO here.

(1) Monthly Abstract of Statistics : It presents data about different facets of Indian economy such as (i) Labour and Employment, (ii) Fuel and power, (iii) Minerals, (iv) Industrial production, (v) Transport, (vi) Foreign trade, (vii) Banking and currency, (viii) Consumption and stocks, (ix) postal traffic, (x) prices etc.

(2) Monthly statistics of the production of selected industries. It gives the monthly production statistics. Two months statistics are published in a combined issue, covering a large number of items. It also contains the index of industrial production and mill stock

position. Separate production figures for different states are presented in it.

(3) Annual Survey of Industries : The National Sample Survey Organization, carries out annually a survey of industries known as Annual Survey of Industries (ASI). Industry-wise estimates of employment, output, input and capital are displayed separately for the census and sample sectors. Because of the time lag in the availability of results of ASI, the net value added estimates for more recently years are prepared from summary result of Factory Sector, containing combined results of census and sample sectors. The report is brought out within a year of the completion of the survey. Data with regard to defence are kept out of the purview of this survey as they are not covered under ASI. This is an annual publication till 1958, it was published under the name 'Census of Indian Manufactures'.

(4) Statistical Abstract of India : This publication is of general nature. It gives an account of area, population, climate, agriculture, mining, banks, motor vehicles, balance of payments etc. Large number of tables are presented relating to social, economic and natural factors.

(5) National Accounts Statistics (NAS): This annual publication is also known as white paper. It incorporates the estimates of capital, formation, saving, private consumption expenditure, factor incomes and the disaggregated tables. The white paper published in January, 1978 introduced for the first time, a more recent base 1970-71 for the series at constant prices replacing the earlier base 1960-61.

The issue of NAS published in April, 1980 contains data up to 1977-78.

There are many other publications of CSO, which are referred ahead in this chapter.

NATIONAL SAMPLE SURVEY ORGANISATION (NSSO)

Census studies have been very expensive and time consuming. In many situations, we cannot afford to wait for long, to get the overall picture of various aspects of socio-economic study. Hence, the emphasis was laid down to Sampling Studies which led to the creation of National Sample Survey Organisation. As given earlier, it was established in January, 1950 in the Department of Economic Affairs, Ministry of Finance, to conduct countrywide multipurpose sample surveys, covering all aspects of national economy required by National Income Committee (NIC), Planning Commission and other ministries of Government of India. The Directorate of National Sample Survey, used to work under the statistics Department of the Cabinet Secretariat. Its main functions were :

(i) Collection of socio-economic data relating to demographic conditions for the whole country on regular basis.

(ii) To provide statistical data for national income and planning.

(iii) To conduct annual surveys in the organised industrial sector.

(iv) Training of personnel and providing guidance to the states in the conduct of surveys.

The work of NSSO was being done under dual control for two decades. The field work was carried out by the Directorate of NSS, whereas designing of surveys, processing of data and preparation of reports was entrusted to the Indian Statistical Institute. In March 1970, the Government of India accepted the recommendation that all aspects of the NSS be brought under an unified control, with a Governing Council to govern its activities. The council was given the requisite independence and autonomy in making decision. In the real sense, it materialised in 1972. Now NSSO is under the Department of Statistics, Ministry of Planning, Government of India. The objective of the NSSO, as laid down in the background of the aims of government statistical work, an autonomous organization, are :

(i) To provide statistical and other information needed for the efficient conduct of government business.

(ii) To evolve statistical technique to bear on the analysis of information, the solution of administrative problems and the estimation of future trends.

(iii) To collect and publish information which will be of use to those engaged in economic activities in the country.

(iv) To provide and analyse information which are useful to the research workers.

(v) To assist in keeping the public informed of the new developments in economic and the social fields.

In short, NSSO has unified control of the governing council with regard to survey designs, field operations, data processing, economic analysis and publication of data. The present structure of NSSO consists of four functional divisions, with a Chief Executive Officer at the apex.

POPULATION STATISTICS

Population statistics is the oldest in the history of statistics, collected and compiled by any kingdom or country. Population statistics at that time, used to mean only the counting of heads which was necessary to know the manpower of a kingdom, to protect oneself from other kingdoms or to win the other kingdoms. Now a days, population statistics is a part of demography science which covers genealogy, human agencies, population census, vital statistics and statistical pathology. Here we shall briefly discuss the collection of data pertaining to the population census and the vital statistics only.

Population Census : Enumeration of individuals under state direction at regular intervals is called population census.

Vital Statistics : It is a branch of statistics, which deals with the registration of facts concerning birth, marriage, divorce, sickness and death under the state direction.

Methods of Conducting Population Census

Population census in India, started in 1881 for the first time. Before independence of India, census was not conducted under a permanent legislation. But after independence, Government of India passed an act known as the Indian Census Act, 1948 which empowers the central government to take a census, appoint necessary census staff, delegate authority and the status of public servants on census officer, assume the power to call upon persons to give assistance and any person is legally bound to answer any question asked of him by a census officer for the census information. It also authorises the census enumerators to enter a house and interrogate people. Any defiance to this rule shall be punishable. After independence, there has been four censuses, the last was conducted in 1981. In fact, two methods of census are there.

(1) *De facto* method or data system.
(2) *De jure* method or period system.

Both these methods are further sub-divided into canvassers method and household methods.

De facto Method

In this method, persons are counted, wherever they are, on the data of census. In this way, counting takes place on one date and that is the reason it is called the date system as well. This gives good information about the number of people in a country but suffers from many lacunae. Firstly, it excludes people where enumerators cannot reach. Moreover, it involves a very huge number of enumerators and hence this method is uneconomical.

De jure Method

In this method, people are counted on the basis of their normal (habitual) residence. Any person found temporarily at a place is excluded from enumerator and counted by an enumeration at his place of residence, even if he is a absent. In this method, census is not conducted on a fixed date but is conducted in a particular period. That is why, it is known as a period system. The *de jure* canvasser method provides geographical and regional distribution of population more exactly than *de facto* method. The period of census under the *de jure* method since 1941 has been extended to three weeks.

Slip System Instead of Filling the Schedules

The slip system of census was introduced for the first time in India during the census of 1941. In this system, one slip was assigned to each individual on which the entire information about the person concerned was noted. The slip system is very convenient for the compilation and the analysis of data. Answers to various questions were recorded in the form of symbol codes on the slip. This saves a lot of labour and time of the enumerator and eases the process of compilation of information, with the help of mechanical tabulation.

Census 1951, the first census of free India, was conducted in 1951. The Census Act of India 1948, was passed for the conduct of the census. The office of the census commissioner was created on a permanent basis. Otherwise, this office used to be wound up when the census operations were over. The full designation of the post was that of a Census Commissioner and Registrar General of India. The officer worked as a Census Commissioner during the period of census and carried on the duties of Registrar General during intercensal period.

It was for the first time in 1951, that a distinction between House and Household was made. House was defined as a dwelling place with a separate main entrance. Household was defined on the basis of *chulha* i.e., persons or group of persons living together and dining from a common kitchen were regarded as the member of the same household. This helped in estimating the family size. Enumeration of population was conducted on the basis of household in 1951 census.

A national register of citizens had been prepared for the first time, which served as an inventory for various official and non-official investigations. This register had been prepared for every village and every town. The Registrar General keeps this register up-to-date, by recording births and deaths. The Registrar General is required to keep a copy of the relevant portion at the village block and district headquarters. This register is also used in preparing the electoral lists.

To acquire more information about the economic characteristics of the population, the changes were made in the existing slip of 1941 for the census in 1951.

Census 1961 *de jure* census method was adopted at this time and hence, each individual was contacted at his place of residence. Some improvements were made in 1961 census over the census 1951. The changes are mentioned in brief. Each house was identified by a location code. In census 1961, three categories were defined namely, (i) Building, (ii) House, (iii) Household. The word building referred to an entire structure raised on the ground whereas, a census house may be whole or a part of a building. Hence, a building can have or more census house.

The questionnaire for 1961 census was prepared in consultation with the Planning Commission. The questionnaire consisted of two parts namely, (i) household schedule and (ii) the individual census slip. Besides these, a special form was prepared to collect information from technical diploma or degree holders.

A new occupational classification was used in this census. The whole population was divided into two categories namely, working and non-working.

The 1961 census covered Jammu and Kashmir and other snow bound areas for the first time. The first publication of the census 1961 results for the whole country appeared on March 25, 1961 i.e., just after three weeks of the enumeration work. Census 1971, the preparation for census 1971 started as early as 1967. A seminar was held in New Delhi from the 3rd to 6th May, 1967, for evolving a procedure and preparing schedules and questionnaire to remove the deficiencies of the census of 1961 and to obtain any other required information. After the seminar, a questionnaire was prepared and tested in the field. The experiences of pre-testing were discussed again in the conference of superintendents of the census operations of the states at Delhi in January 1968. The enumeration work of census was carried out from the 10th March to the 31st March, 1971. The enumerators revisited their area from 1st to 3rd April, to make the information more complete with regard to the birth and deaths that occurred before the sunrise of 1st April 1971. Also, on the night of 31st March 1971 each enumerator counted the number of homeless in their areas of survey.

The special features of 1971 census were, (i) the data were collected on current fertility, (ii) a separate card known as Degree holders and technical personnel card was filled in the 1971 census, (iii) for the first time, detailed information was collected on migration, (iv) for the first time, census data computerised.

Economic Census

The Central Statistical Organization had taken a major step during 1977, in collaboration with the state statistical bureaus to remove data gaps in the unorganized sectors of non-agricultural economy, covering unregistered manufacturing, trade, transport and services, by launching a central plan scheme on economic census and surveys. This economic survey of non-agricultural establishments, employing at least one hired person was complete in the last quarter of 1977. The information regarding the type of activity, number of working persons and type of ownership etc., were collected and used in the sample survey of selected establishments in different sectors, conducted during 1978-79 and 1979-80. These surveys collected information relating to the quantity of inputs, outputs, investments, emoluments, employments etc. The manufacturing sector was covered in the 33rd round of NSS during 1978-79 and the sectors of the trade, transport, hotel and restaurants, storage and warehousing were covered in the 34th round of NSS, conducted in 1979-80. The small establishments were also covered in the 33rd and the 34th rounds.

The census covered large number of small establishments, which were subject to high morbidity and mortality and the variation in their input and output. Hence, the need was felt for conducting the economic census at regular intervals. Five year interval, for such a census was considered, as such, the second census was due in 1982. But the preparations were going on in 1980 for census 1981, particularly with regard to the preparation of house lists. Hence, it was considered beneficial that the second economic census be synchronised with the house listing operation meant for the 1981 population census. Economic census of 1980 may be considered a by-product of the 1981 population census.

Individual slips from sampled blocks were also filled by the enumerators for specific information. Separate cards were filled by persons holding a recognised degree or diploma in science, engineering, technology or medicine etc. This card contained information about his bio-data and employment.

Compiled census data have been published for India, as a whole, and in parts for each state, district in the form of census reports. Various aspects covered during census have been tabulated and published systematically.

Agricultural Statistics

India is an agricultural country and its economy largely depends on agriculture. Since ancient days, agriculture had been the main source of income of all the rulers. They collected money in the form of land revenue. Therefore, most of the rulers mentioned records of agricultural land and their categories on the basis of fertility and production. Records were also maintained for orchards. The main objectives of keeping these records was the collection of revenue.

After independence, a lot of emphasis has been given to agriculture. Indian economy is based mostly on agriculture even today, hence, a comprehensive and reliable statistics for good planning and development of agriculture and vis-a-vis of the country are necessary. As a matter of fact, all statistics which have an impact on agricultural economy may be regarded as agricultural statistics. Here, we shall use the term agricultural statistics in a narrower sense. The statistics pertaining to land utilization, production of crops, livestock and agricultural prices, poultry, forestry etc. will come under the category of agricultural statistics.

Agricultural statistics is mostly collected by the Directorate of Economics and Statistics (DES) at the centre as well at the state level. Each state has a Directorate of Economics and Statistics. The statistics regarding area, crop production, poultry and forests are collected. Total area statistics are obtained from two sources (i) the Surveyor General of India, (ii) the village records maintained by the Revenue Department. The area is further classified as forest, land put to non-agricultural uses, barren and unculturable and, permanent pastures and other grazing lands, culturable waste, current fallows, other fallow land, new area sown (irrigated and unirrigated). The irrigated area is found out from the availability of irrigation by canals, tanks and wells.

The area under crops is obtained by two sources (i) official series based on village records, (ii) NSS series based on sample surveys. NSS collects and during the regular rounds of survey on area under different crops. The area figures provided by these two sources are far apart, because of the difference in method of the coverage of crops, difference in the field work, the classification of area under feed and the fodder crops, the allocation of area under mixed crops and due to sampling error in NSS estimates. The land utilisation statistics are published in *Indian Agricultural Statistics.* The individual states publish these statistics in *Season and Crop Reports.* Land utilisation statistics are also published in (i) *Agricultural Situation in India* (monthly), (ii) *Abstract of Agricultural Statistics* (DES Annual), (iii) *Statistical Abstract of the Indian Union* (CSO, Annual).

Agricultural Production Statistics : Agricultural production includes the production of food and non-food crops but forests, livestock and fisheries are excluded at this place. We will give a brief account of these separately. Crop output statistics cannot be collected at the time of harvesting, as millions of farmers harvest their crops at the time of harvesting, millions of farmers harvest their crops at the same time. Mostly they are uneducated and maintain no record. Hence, the output is estimated by multiplying the area under a crop by the average expected yield per hectare, in the season. The yield statistics are collected by official machinery and NSSO. As official machinery, the work of the survey is undertaken by the Directorate of Economics and Statistics. Two methods are adopted to collect yield statistics of various crops:

(a) the traditional method, (b) the random sampling method.

The traditional Method : This method of estimation is known as Annawari system. In this system, the condition of the crop is judged by Patwari or any other official in relation to the normal crop. This is known as the condition factor or the Annawari estimate. The yield estimate is obtained by the Annawari system, by the formula,

Crop yield=Annas judged ÷ Total Annas × normal yield × area

This system suffers from two main lacunae. One is that the condition factor depends totally on the judgement of Lekhapals or Patwaris, which is very subjective. Secondly, the total annas representing the normal condition of a crop is not uniform in all the states. For instance, some states take fifteen annas as normal condition while Madhya Pradesh takes 13.3 annas to represent normal condition of a crop.

Random Sampling Method : This method was recommended by the Board of Agriculture as early as 1919 but was abandoned due to the huge financial burden on the states. The Indian Council of Agriculture introduced the random sampling method again in 1942, for cotton crops. This method was later extended for other crops in all states. Except West Bengal and Orissa, where the job was entrusted to Indian Statistical Institute, Calcutta. In random sampling method, villages were randomly selected in tehsils of a state and within each village few fields were selected randomly. To be more specific stratified multi-stage random sampling design is adopted for the crop cutting experiments. Crop cutting experiments have been conducted since 1943. Now the experiments are conducted by the Directorate of Economics and Statistics and statistical section of the Board of Revenue, through its field staff and separately by National Sample Survey Organisation. The NSSO conducts crop cutting experiments on major cereal crops, during the course of their regular survey rounds. The yield data collected by two agencies differ in their estimates quite widely due to the difference in the shape and the size of the cuts, the difference in drainage factors, the difficulties of harvesting at a proper time by NSSO investigators, the difference in the methods of estimation and overall difference in the field work.

The results of the survey are published mainly in (i) the *Agricultural Situation in India* (monthly), (ii) the weekly *Bulletin of Agricultural Prices,* (iii) the *Agricultural Statistics.* A long list of other publications providing statistics related to Agriculture can be seen on the backside of the journal, *Agricultural Situation in India.*

A detailed Agricultural Census in India was conducted in the year 1970 -71 from July 1, 1970 to 30th June, 1971. In this survey, information was collected about holding, land -use under different crops (irrigated and unirrigated), water source-wise irrigated area etc. The second Agricultural census was conducted during the agricultural year 1976 -77 i.e. from July 1, 1976 to June 30, 1977. This covers one *kharif* and one *rabi* season completely. In this census, additional information was collected from the land holders of 2 per cent villages about the use and consumption of fertilizers, insecticides, livestock, farm implements, saline and alkaline lands, in addition to the information, which was collected in 1970-71 Agricultural Census.

Agricultural Census 1980-81 is third of its kind in India, sponsored by United Nations Food and Agricultural Organisation (FAO). In India, it is being conducted under the patronage of the Ministry of Agriculture in the centre and Department of Agriculture, District Statistical Officer and other government officials actively participated to get the enumeration work done properly. This census was conducted into phases, as was done in 1976-77 census. In the first phase, enumeration work had been completed and in the second phase, the information was collected from land holders as per revenue records. The work of the second phase was carried out by the revenue Department. The results of the survey are under the process of publication.

Another agricultural survey named, the Input Survey, was conducted during the agricultural year 1981-82 i.e. from 1st July 1981 to 30th June, 1982. The input survey was conducted in about 7 per cent of inhibited villages. Institutional holdings were kept out of the sampling frame. City people were also not included in the input survey. For the selection of land holdings from a selected village, the land holders were divided into five groups i.e. farmers with the holding less than 1 hectare, 1-2, 2-4, 4-10 and more than

10 hectares. From each group of farm holding, 4 families were selected randomly. If any group contained 4 or less than four farmers, all were included in the sample. The enumerators collected the information in a single visit for both the seasons i.e. *kharif* and *rabi.* Information was collected about the individual and joint holdings.

The enumerators collected information about (1) the number of plots, (2) the multiple cropping pattern in irrigated and unirrigated areas, (3) the saline, alkaline and water logged areas, (4) the application of chemical fertilizers, manures and insecticides in main crops (separately for irrigated and unirrigated sown areas), (5) livestock, (6) agricultural implements, and (7) agricultural credits.

The results of the survey are published in the form of a report, by the Directorate of Economics and Statistics, *Output from Agricultural* for the purpose of evaluation of output, 68 agricultural crops, including by-products are considered. These are divided into four broad categories viz. (i) 35 items of principal crops, (ii) 10 items of minor crops, (iii) 11 items of miscellaneous and suspected crops.

Data on area under unspecified crops are available annually in *Indian Agricultural Statistics,* published by DES with a time lag of two to three years. Some other publications (regular and *ad hoc*) giving Agricultural Statistics are as given below :

1. Bulletin of Food Statistics (DES Annual)
2. Oilseeds in India (DEZS, 1955, 1961)
3. Bulletin on Commercial Crop Statistics (DES).
4. Farm Management/Cost of Cultivation Studies (DES, *ad hoc*).
5. NSS Report No.- 32 (Part I to III), 140.

Livestock and Poultry Statistics

On the initiative of the Government, the first cattle census, on all-India basis, was conducted from December 1919 to April 1920. Since then the livestock census is held quinquennially by the Directorate of Economics and Statistics. Statistics regarding livestock and poultry are published in *Indian Livestock Census* (quinquennial). This publication contains information related to cattle and poultry.

For the purpose of the estimation of the gross value of output from livestock and poultry, seven broad groups are taken into consideration, namely, (i) milk and milk products, (ii) meat and meat products, (iii) hides and skins, (iv) eggs and poultry meat, (v) wool and hair, (vi) dung and other products.

Industrial Statistics

This is an industrial age. India is making her best efforts in the direction of industrialisation. As such India had come under the category of industrially advance countries. There is a need to collect and publish industrial statistics. For the purpose of domestic product estimation, the entire manufacturing industry is subdivided into two broad sectors viz. manufacturing registered and manufacturing unregistered. The registered manufacturing sector covers all organised manufacturing and processing establishments (classified as factories) which are registered under Indian Factories Act (IFA), 1948. Unregistered manufacturing sector covers all manufacturing units which are not registered under the IFA, 1948. It covers all manufacturing units employing less than 10 workers if using power, or less than 20 workers if not using power.

The history of industrial statistics begins from the year 1942. When in Industrial Statistical Act, 1942 was passed. This act empowered the state governments to collect data from factories, with regard to any statistical information. A Directorate of Industrial Statistics was set up at the centre to enforce this Act. The first census of manufacturing industries was conducted in 1946 and the results were published in a bound volume. From 1959, the census of the manufacturing industries and sample survey of industries was discontinued and the Annual Survey of Industries (ASI) was started. The NSSO carries out the annual survey of industries regularly. This survey provides industry-wise estimates of employment, output, input and capital employed separately for the census and sample sectors Data in respect of Defence production are collected independently as they are not covered under ASI. The result of annual survey of the census sector are published in bound volumes by CSO and the results of the sample sector are published by NSSO in their regular reports.

As regards statistics of unregistered manufacturing units, NSSO conducts surveys on small scale manufacturers and self employed households, in non-agricultural enterprises. NSSO conducted surveys in the 23rd and the 29th rounds respectively. The 23rd round included the small scale manufacturing in rural and urban areas in 1968-69. The results of the survey are presented in NSS Report No. 205. The 29th round was the survey of self-employed households in non-agricultural enterprises in 1974-75. Tables with notes are published in NSS Report No. 280/6.

The NSS surveys only cover effectively the household establishments and do not provide information on the value added per worker for the non-household enterprises. However, data are available from the survey results conducted under the centrally sponsored scheme on the Survey of Small Scale Industries by the Ministry of Industrial Development.

The remaining factories constitute the sample sector 1970-71 as the reference period. All India Report the small scale industrial units, covering mainly urban areas, was published in 1977 by Development Commissioner Small Scale Industries (DCSSI).

Different publications for Industrial Statistics are :

1. *Monthly Statistics of the Production of Selected Industries* : It is a publication of the Industrial Statistics wing of the CSO.

2. *Monthly Abstract of Statistics (CSO)* : It contains data pertaining to industrial production, index number of industrial production, consumption and stocks etc.

3. *Statistical Abstract of the Indian Union* (CSO, Annual) : It covers data on various aspects of industry and power. Statewise figures are provided for the latest year. Factories employing 50 or more workers with power or 100 more workers without power constitute the census sector.

4. *Monthly Jute Bulletin* : It is a publication of the Indian Central Jute consumption and export of jute goods etc.

5. *Monthly Statistical Bulletin*, Indian Cotton Textile Industry.

6. *Indian Sugar* : It is published by the Sugar Manufacturers Association.

Besides the above publications, the number of other Reports, Bulletins and Journals are regularly published by various departments, statistical Bureaus, Directorates, Associations etc.

Trade Statistics

Trade statistics is compiled and published regularly by the Directorate of Commercial Intelligence and Statistics (DCIS), Calcutta. The data for trade statistics emerge as a by-product of official administrative data of the nation. These statistics are given with regard to, (i) inland wholesale and retail trade, (ii) foreign trade of India, (iii) index number of inland and foreign trade. The Directorate of Commercial Intelligence and Statistics started the publication of Indian Trade Journal in 1906. Different publications which give data on trade are :

1. Accounts relating to the Inland (Rail and River borne) Trade of India (DCIS, Quarterly). The journal contains data as provided by railways and steamer authorities to whom the traders have to submit their invoices.
2. Statistics relating to the coasting trade of India are published in the Bulletins :

   (a) Statistics of Foreign and Coastal Cargo Movements of India (DCIS, Monthly).

   (b) Statistics of coasting trade of India (DCIS, Annual).

   (c) Statistics of Maritime Navigation of India (DCIS) : It contains data available with the DGCIS. Coasting trade is registered separately by the traders in custom houses.
3. *Indian Trade Journal* : (DCIS, weekly). It contains some data about retail trade of India.

Some publications which supply compiled information about foreign trade are :

1. (a) *Monthly Statistics of Foreign Trade of India,* Vol. I & II (DCIS) : Volume I contains information related to the exports and re-exports. Volume II contains data about the imports. This publication uses the trade classification based on standard International trade classification.

(b) *Supplement to the Monthly Statistics by the Foreign Trade of India,* Vol I & II (DCIS, Quarterly) : This publication contains mostly the information which has already been published in a monthly series.

2. *Annual Statistics of Foreign Trade of India by Countries* : This issue contains totals of the monthly statistics, with certain revision and corrections. These revisions and corrections arise due to short and shut-out shipment in exports and finalisation of note pass transactions in imports.

Labour Statistics

The term labour refers to the workers is characterised largely by physical exertion. Generally, they are the persons engaged wholly or partly in the production of goods and services. These persons may be engaged in agriculture and allied activities, forestry and logging, fishing, mining and quarrying, manufacturing construction, transport, storage, communication, trade, hotel and management banking and insurance, education and research, health and medicines etc. But the term labour is mainly in connotation to the workers in industry, agriculture and construction. Out of the three, main emphasis is laid on industrial labour. International Labour Organisation (ILO) is playing a significant role in the collection and publications of labour statistics. The main function of ILO is to maintain uniformity of labour statistics in all the countries. This makes comparison feasible on an international level.

The labour statistics is collected and compiled relating to average daily employments in Factories, Mines, Plantation, Railways, Post and Telegraph, Commercial Establishments and

Central Government Establishments, wages and hours of work, levels of living, family living, statistics of accidents and occupational diseases, statistics of trade unions and disputes, social security, migration etc.

There are three main sources of employment statistics in India namely, (i) Decennial Population Census, (ii) National Sample Survey, (iii) Employment exchanges. The statistics of wages are the indicator of the economic condition of the fixed salaried people. Level of living is adjudged in terms of the wages paid to the workers and the consumer price index, at a time. The Minimum Wage Act, 1948 provides the protection to the labourers against price rise of essential commodities and services.

Statistics of accidents and occupational diseases are collected under the Workmen's Compensation Act of 1923. The figures for the number of injuries and accidents, for which the compensation is paid, are collected. The accidents include date relating to deaths, permanent disablement and temporary disablement. Social security is provided to the employees getting less than Rs. 500 per month, under the Employees State Insurance Act of 1948, in all the registered factories. Benefits under this scheme accrue to the employees with regard to sickness, maternity, disablement and dependents.

There are three organisations which are mainly responsible for the collection and publication of labour statistics. They are given below with a brief description of their functioning.

Labour Bureau : It comes under the Ministry of Labour, Employment and Rehabilitation, Government of India. Labour Bureau is main source of labour statistics, as it publishes a number of journals, books and *ad hoc* report. Its main publications are :

(i) *Indian Labour Journal* (Monthly)
(ii) *Indian Labour Year Book* (Annual)
(iii) *Indian Labour Statistics* (Annual)
(iv) *Trade Union in India* (Biennial)
(v) *Statistics of Factories* (Annual)
(vi) *Pocket Book of Labour Statistics* (Annual)

These publications contain information about the items given earlier.

Directorate of Employment and Training (DET) : This Directorate is located in New Delhi. The Director General collects and analyses the data on the occupational pattern of the employees. The employment exchanges are the main source of data. The data are published in the form of journals and reports.

(i) *Quarterly Employment Review* : The journal contains data of the employment in public and private sectors, industrywise, statewise and zonewise. It also gives information relating to women employees.
(ii) *Area Employment Market Report* (Quarterly and Annual)
(iii) *State Employment Reports* (Quarterly)
(iv) *Reports on Occupational Pattern of Employees* (Biennial)
(v) *Reports on Biennial Survey of Establishments* employing workers in private sectors (Biennial).

Director General of Mines Safety : The Directorate of safety, collects data relating to labour employed in collieries, their wages and hours of work, productivity, general health, accidents etc. The information about all aspects is published in the Monthly Coal Bulletin.

Statistics of Agricultural Labour : The statistical information regarding agricultural labour has been collected through Labour Enquiry Committees, appointed from time to time. The first enquiry was conducted in 1950-51, under the supervision of Directorate of Economics and Statistics. The second enquiry was conducted in 1956-57, under the Ministry of Labour in collaboration with CSO, NSSO and ASI. In both these surveys only landless agricultural households were selected for investigation. The Directorate of Economics and Statistics collects information about Agricultural Labour (skilled and unskilled) regularly. Collected statistics are published in the following two journals :

(i) *Agricultural Wages in India.*
(ii) *Agricultural Situation in India.*

National Income Statistics

National income estimates are of great importance, to get a broad view of the entire economy of a country. It also provides information about the changes occurring in economy from year to year. The economic policies and planning, are mostly framed keeping in view, the national income of a nation. The National Income Committee (NIC), set up by the Government of India in 1949, produced for the first time, national income estimates for the entire Indian union. The estimates, along with the mythology, were published in the first and final report and final report of the National Income Committees (Ministry of Finance) in 1951 and 1954. Efforts were made to improve the empirical data and overcome the gaps. The first results of these efforts were presented by the CSO in the National Income Statistics. The result of the earlier studies along with the revised series of national income were published in Brochure on the Revised Series of National product for 1960-61 to 1964-65 by CSO in 1967. Two more Brochures were published by the CSO.

(i) *National Income Statistics :* Estimates of Capital Formation in India from 1960-61 to 1965-66 were presented in 1969.

(ii) *National Income Statistics :* Estimates of Saving in India from 1960-61 to 1965-66 were released in 1969.

Here, it would be worthwhile to give some idea about the national income of a nation. Different economists defined national income in various ways without entangling with these controversies, we give the definition : Sum of the employees compensation and the net income from property and entrepreneurship, that is the distributed factors income, represents the national income of the country.

The national income becomes gross, when added with consumption of fixed capital. By definition, it is equivalent to the national product at factor cost, gross and net.

In simple words, national income may be defined as, the value of commodities and services produced during a particular period counted without duplication.

Income attributable to factor services rendered by the normal residents of the country, to the rest of the world, less factor services rendered to them by the rest of the world. It also includes retained earnings of foreign controlled rupee companies, and branches of foreign companies in the domestic territory. Residents include both individuals and institutions. Tourists or commercial travellers of a given country travelling abroad are also treated as residents. The factor income generated by diplomats and counsellors representatives and members of armed forces stationed abroad are the domestic product of the resident country. Factor incomes of locally recruited staff of the foreign diplomatic military establishments are included in the factor income from abroad.

Uses of National Income

1. The first and foremost use of national income estimate that it gives us a correct picture of the structure of the economy of the nation, as well as the distribution of income according to regions, industrial origin and income from functional services and persons during a specified period of time.
2. National income estimates give us an idea of the purchasing power of the people in the country. Inflationary and deflationary gaps of purchasing power are revealed by income and product figures. This is of inflationary and deflationary programmes.
3. National income statistics provide a useful guideline in the formulation of the budget of a country.
4. National income figures help forecast the level of business activity for months and years ahead, since it enables us to compare and analyse changes in the output, income and expenditure of the various sectors, as well as entire economy over a period of time. These figures are absolutely fundamental to the analysis of the overall economic and business situations, particularly with regard to its cyclical aspects.
5. National income figures have been found useful for studies of the problems of the economically underdeveloped countries.
6. National income statistics help in making the inter-regional and inter-state comparisons of national income distribution. It is also useful in comparing the economic conditions of a country over a period of time.

7. National income statistics, gives us an idea about the standard of living of the nationals of a country. If we consider the national income only on current prices, the per capita income may increase, but it may not lead to the income in standard of living since the increase in prices causes the fall in the real income of the consumers. To estimate the real national increase, at present the base year is 1970-71.
8. National income estimates provide a basis for the future planning of a country. In the absence of such statistics, no comprehensive economic plan can be chalked out. It also helps in evaluating the success of an economic plan.

Methods of National Income Estimation : The methods to estimate the national income of a country necessarily depend upon the availability of statistics. Commonly, the following three methods are used for the calculation of national income.

(1) Output or production method.
(2) Income method.
(3) Expenditure method.

Output Method : Output or production method is also called inventory method. It consists of finding out the market value of all goods and services produced by the individuals and business enterprises during a specified period of time. The national income or net domestic product can be calculated by the following equation.

National Income = (value of goods and services + Self consumption + increase in stock — depreciation of capital — income from abroad)

The above sum of income contributes the national income at the factor cost.

Income Method : The income method consists of adding together all incomes by the way of wages, interests, rents and profits. Wages include the value of services (mental and physical), in terms of money rendered by the nationals. Interest is the amount of money, received by the entrepreneurs, by way of capital investment. Similarly, rent may be defined as the factor income generated by the letting and the use of land for agricultural and other purpose, buildings (residential and non-residential), machinery equipment and other fixed assets. It is treated as rental income from property. Also any amount of money saved by the individuals or institutions, after deducting the expenditure on production, may be termed as profit.

Expenditure Method : The national income is equal to the sum of consumption and saving. Symbolically, we can write

$$Y = C + S$$

where, Y = National income
C = Consumption
S = Saving

The method is rarely used because of the non-availability of data required for its calculation. Because of the predominance of unorganized sectors in under-developed and developing countries, there is a lack of proper maintenance of records of consumption, expenditure, saving etc.

As a matter of fact, all the three methods lead to the same value of national income estimates. These are three angles from which we can look at the national income.

Problems in the Estimation of National Income

Every country, advanced or underdeveloped, faces some difficulty in computing national income. However, underdeveloped countries are faced with some special difficulties arising out of the peculiar nature of such economics. In our country, the important problems that are faced while estimating national income are :

1. There is a substantial non-monetised sector which makes the calculation of national income difficult. A great deal of what is produced in other subsistence sector is either exchanged for other goods or is kept for personal consumption. This tends to understate the national income.
2. In underdeveloped countries due to the lack of occupational specialisation, calculation of national income by industrial origin becomes rather difficult. For instance, besides the

crops, farmers often produce a variety of products like—eggs, mild, articles of clothing etc., that are not included in the calculation of national income estimates due to the lack of the availability of such data.

3. In developing countries, people are mostly illiterate and do not keep any accounts and even if they do, they are reluctant to disclose their income correctly. In such a situation only rough estimates are possible.
4. The computation of national income in terms of money, underestimates the real income, because it does not include the real cost of the produced articles. The income earned by two people may be the same, but if one works for longer hours than the other, there is some justification in saying that the real income of the former is lower than the latter.
5. National income estimates fail to measure adequately, the changes output due to the changes in the price level. Index numbers, used to measure changes in the price level are simply rough estimates. Therefore, the national figures under such circumstances are often misleading.
6. National income estimates include only those goods and services, which are commercially used. People living in rural areas usually manufacture articles of consumption, from traditional goods/raw material, are able to avoid many expenses. For example, they build their own huts, garments and other items of necessity. The value of such goods are not included in the national income estimates.
7. The international comparison of national income is inaccurate and misleading due to the exchange rate conversion of different countries into a common currency i.e., U.S. dollars. The rates of exchanges are primarily based on the prices of internationally traded goods. Unlike developed countries, in underdeveloped countries there are many goods and services that are never traded internationally and are also priced low. This makes international comparisons of national incomes misleading and superfluous.
8. While computing national income of a country, the problem of double counting may crop up. Double counting refers to the possibility of a commodity or raw material or labour being

included in national income calculation more than once. An example will elucidate this point. Suppose we calculate the total money value of a book published in a particular year and include it as part of the national income of the country for that year and suppose at the same time, we calculate the money value of paper used in the printing of books and add it up to the national income of the year. The value of paper has entered the national income calculation twice—once directly and second time indirectly through the value of the books. This is a problem of double counting. The best way to avoid this duplication error is to calculate only the value of those goods and services that enter into the final consumption stage.

Financial Statistics

Financial statistics are the best source of understanding the state of economy of a country. These statistics contain compiled data relating to Banking, Currency, Exchanges and Bullion, Public Finances, Statistics of Insurance Companies and Financial Institutions are also the part of financial statistics.

The Reserve Bank of India (RBI) has a full-fledged department of research and statistics. This department compiles data and publishes information on banking, exchange, currency, finances etc. These data are used by the Government of India, Foreign Governments and International Monetary Fund (IMF) and private sectors. Some important publications are :

1. *Reserve Bank of India Bulletin (Monthly) :* Besides other things, it contains data on banking money and credit.
2. *Statistical Tables Relating to Banks in India (RBI Annual) :* These tables furnish data on scheduled and non-scheduled Banks, whether registered in India or abroad.
3. *Report on Currency and Finance (RBI, Annual) :* This report has two parts. Part I covers information about investment, finance and advances for agriculture, external assistance, international monetary situation etc. Part II gives details in respect of price trends, rural credit, capital market, trade, tariffs and exchange control, world trade, currency, coinage etc.

4. *Report on the Trend and Progress of Banking in India (RBI, Annual)* : This publication gives data relating to liabilities and assets of the RBI, consolidated position of other scheduled banks, position of investment of banks, interest, money rate etc.
5. *Statistical Statements Relating to Cooperative Movement in India Vol. I & II (RBI, Annual)* : This publication contains data relating to provincial and central cooperative bank, credit societies, land mortgage banks and other types of cooperative societies.
6. *Banking Statistics* : Basic Statistical Returns (RBI, Annual).
7. *Reserve Bank of India* (Annual Report).
8. Analysis of the Accounts of the Joint Stock Companies (RBS Bulletin).
9. Balance of Payment Statistics (RBI, Bulletin).
10. Results of Sample Studies on Analysis of Finance of Medium and Large Public Companies, (RBI, Bulletin).
11. Survey of Finances of Local Authorities, (RBI, Bulletin).
12. Study of Company Finance (RBI, Bulletin).
13. Annual Report and Accounts (L.I.C. of India).
14. Abstract and Statements in respect of Valuation of Assets and Liabilities (L.I.C. of India).
15. Finance Accounts of the Central and State Governments (Ministry of Finance, Annual).

Besides the above publications there are many more as Annual Reports, Regular Journals, Brochures, Bulletins etc., which furnish financial statistics. All are not listed here. Financial statistics also take into account, a number of sectors which are not covered in the discussion here.

The statistical organization in India is very wide. It covers hundreds of sectors which are functionally important and contribute a lot to the private and public sector. In this chapter, a brief description of some important organisations have been given. Some of the uncovered sectors are Trade, Hotel and Restaurants, Transport, Storage and Communication. Anyhow, the information given in this chapter is enough to obtain any allied information as well. The names of other publications will become available in the publications named in this chapter.

Bibliography

Alvi, Zamir (1995), *Statistical Geography—Methods and Applications*, Rawat Pub., Jaipur.

Anderson, T. (1956), Potential Model and Spatial Distribution of Population, *Papers and Proceedings of Regional Science Association*, Vol. 2.

Ahmad Qazi (1965), *Indian Cities : Characteristics and Correlates, Research Paper No. 102*, Deptt. of Geography, Chicago University, Illinois.

Anderson, T.W. (1958), *Introduction to Multivariate Statistical Methods*, John Wiley & Sons, New York.

Bhatia, S.S. (1965), Pattern of Crop Concentration and Diversification in India, *Economic Geography*, Vol. XL, pp. 38-56.

Berry, B.L.J. (1970), An Inductive Approach to the Regionalization of Economic Development in Ginsburg (ed.), *Essays on Geography and Economic Development*, Chicago University, pp. 78-100.

Biplab Dasgupta (1971), Socio-Economic Classification of Indian Districts—A Statistical Approach, *Economic and Political Weekly*, New Delhi.

Birch, J.W. (1960), A note on the Sample Farm Survey and its Use as a Basis of Generalized Mapping, *Economic Geography*, Vol. 36, p. 254.

Berry, B.J.L. (1964), Approaches to Regional Analysis—A Synopsis, *Annals of the Association of American Geographers*, Vol. 54, p. 2.

Blommers, P. and Lindquist, E.F. (1960), *Elementary Statistical Methods*, Oxford Book Company.

Board, C.; Chorley, R.J.; and Stoddart, D.R. (eds.), (1969), *Progress in Geography*, Vol. 1 to 7, Butler and Tanner Ltd.

Brian, J.L.; Berry, B.J.L.; and Marble, D.F. (eds.), (1968), *Spatial Analysis—A Reader in Statistical Geography*, Prentice Hall, Inc.

Berdie Douglas R., and Anderson, John, F., (1974), *Questionnaire Design and Use*, Metuchen, N.J. The Scarecrow Press, Inc.

Boot, John, G.C. & Cox Edwin B. (1979), *Statistical Analysis for Managerial Decisions*, 2nd. ed. New Delhi, McGraw Hill Pub. Co. Ltd., International Student Edition.

Bowley, A.L., (1937), *Elements of Statistics*, 6th Ed., London.

Clark, P.J. and Evans, F.C., (1954), Distance to Nearest Neighbour as a Measure of Spatial Relationships in Population, *Ecology*, 35, pp. 445-53.

Cooley, W.W. and Lohnes, P.R., (1971), *Multivariate Data Analysis*, John Wiley & Sons, New York.

Chance, William, A., (1975), *Statistical Methods for Decision Making*, Bombay, D.B. Tarapoorevala Sons & Co. Pvt. Ltd.

Chaturvedi, J.C. (1953), *Mathematical Statistics*, Agra.

Child Dennis, (1973), *The Essentials of Factor Analysis*, New York.

Cochran, W.G., (1963), *Sampling Techniques*, 2nd. ed., New York, John Wiley & Sons.

Cooley, William, W., and Lohnes, Paul, R., (1971), *Multivariate Data Analysis*, New York, John Wiley & Sons.

Croxton, F.E.; Cowden, D.J.; and Klein, S., (1975), *Applied General Statistics*, Prentice Hall of India, New Delhi.

Chorley, R.J. and Hagget, P., (1968), *Models in Geography*, Methuen and Co. Ltd.

———— (1969), *Physical and Information Models in Geography*, Methuen and Co. Ltd.

Cochran, W.G., (1962), *Sampling Techniques*, Asia Pub. House, New Delhi.

Deming, W. Edwards, (1960), *Sample Design in Business Research*, John Wiley & Sons, New York.

Davis, I.C.; Macullagh, M., (1975), *Display and Analysis of Spatial Data*, John Wiley and Sons.

Dickinson, G.C., (1963), *Statistical Mapping and the Presentation of Statistics*, Edward Enrold Publisher Ltd.

Doi, K., (1975), The Industrial Structure of Japanese Prefectures, Proceedings of IGU Regional Conference in Japan, The Organising Committee of IGU Regional Conference in Japan, The Science Committee of Japan, Tokyo, 1959, p. 310.

Duncan, D., Cuzzort, R.P., and Duncan, B., (1961), *Statistical Geography* (Problems in Analysing Areal Data), The Free Press of Glencoe.

Dacey, M.F., (1960), A note on the derivation of the nearest neighbour distances, *Journal of Regional Science Association*, Vol. 2, pp. 81-87.

———— (1962), Analysis of Central Place and Point Patterns by a nearest neighbour method, *Land Studies in Geography*, Series B, Vol. 24, pp. 55-75.

———— (1963), Order Nearest Neighbour Statistics for a class of Random Patterns in Multi-dimensional Space, *Annals of Association of American Geographers*, Vol. 53, pp. 505-515.

Dasgupta, B., (1971), Socio-economic Classifications of Indian District—A Statistical Approach, *Economic & Political Weekly*, New Delhi.

Duncan, O.D., (1957), The Measurement of Population Distribution, *Population Studies*, p. 40.

Edwards, Allen, (1967), *Statistical Methods*, Rinehart & Winston, New York.

Ebdon, D., (1977), *Statistics in Geography—A Practical Approach*, Basil Blackwell.

Ferguson, George, A., (1962), *Statistical Analysis in Psychology and Education*, McGraw Hill Book Co., New York.

Fisher, R.A., (1958), *Statistical Methods for Research Workers*, Hafner Pub. Co., New York.

Fruchter, Benjamin, (1954), *Introduction to Factor Analysis*, Princeton.

Finney, D.J., (1965), *An Introduction to Statistical Science in Agriculture*, Asia Publishing House, New Delhi.

Fisher, R.A., (1959), *Statistical Methods and Scientific Inference*, Oliver & Boyd.

Fraser, D.A.S., (1958), *Statistics : An Introduction*, John Wiley & Sons Inc, New Delhi.

Gnanadesikan, R., (1977), *Methods for Statistical Data Analysis of Multivariate Observations*, John Wiley & Sons, New Delhi.

Goodman, R., (1967), *Teach Yourself Statistics*, English Language Book Society and England Universities Press Ltd.

Gregony, S., (1978), *Statistical Methods and the Geographers*, Longman Group Ltd.

———(1969), *Models and Quantitative Techniques in Teaching*, Geography, Vol. 54, p. 5.

Godfrey, Arthur, (1977), *Quantitative Methods for Managers*, Edward Arnold Pub., London.

Green, Paul, E., (1978), *Analysing Multivariate Data*, Dryden Press.

Hagget, P., and Chorley, R.J., (1969), *Network Analysis in Geography*, Edward Arnold Pub. Ltd.

Hagget, P., (1965), *Locational Analysis in Human Geography*, Edward Arnold Pub. Ltd.

Hammond, R. and McCullagh, P.S., (1977), *Quantitative Techniques in Geography* : An Introduction Clarendon Press.

Harnett, Donald L. and Murphy, James L., (1975), *Introductory Statistical Analysis*, Philippines : Addison Wesley Pub. Co. Inc.

Hollander, Myles, and Wolfe Douglas, A., (1973), *Nonparametric Statistical Methods*, New York, John Wiley.

Jacobson, P.E., (1976), *Introduction to Statistical Measures for Social and Behavioural Sciences*, The Dryden Press.

Johnston, R.J., (1978), *Multivariate Statistical Analysis in Geography*, Longman.

John, Peter W.M., (1971), *Statistical Design and Analysis of Experiments*, New York, The MacMillan Co.

Johnson, Rodney D. and Siskin Bernard R., (1977), *Quantitative Techniques for Business Decisions*, Prentice Hall of India, New Delhi.

Johnson, R.J., (1978), *Multi-variate Statistical Analysis in Geography*, Longman.

Kahn, Robert, L. & Charles, F., (1957), *The Dynamics of Interviewing*, New York, John Wiley and Sons.

Karson, Marvin, J., (1987), *Multivariate Statistical Methods*, The Iowa University Press.

Kendall, M.G., (1961), *A Course in Multivariate Analysis*, London, Griffin.

Kish, Leslie, (1965), *Survey Sampling*, New York, John Wiley & Sons.

Kothari, C.R., (1984), *Quantitative Techniques*, Vikas Publishing House Pvt. Ltd.

King, L.J., (1969), *Statistical Analysis in Geography*, Prentice Hall.

———— (1969), The Analysis of Spatial Form and Its Relation tc Geographic Theory, *Annals of the Association of American Geographers*, Vol. 59, p. 573.

Lewis, P., (1977), *Maps and Statistics*, Methuen and Co. Ltd.

Lindley, D.V., and Miller, J.C., (1953), *Cambridge Elementary Statistical Tables*, 3, 5, 7, & 8.

Lewin, Richard, (1979), *Statistics for Management*, Prentiee Hall of India Pvt. Ltd.

Mahmood, A., (1977), *Statistical Methods in Geographical Studies*, Rajesh Publications, New Delhi.

Mandal R. B. (1980), *Statistics for Geographers and Social Scientists*, Concept Publishing Co., New Delhi

Mandal R.B. (1982) *Introduction to Population Analysis*, Concept Publishing Co., New Delhi.

Masuyama, M., (1954), *Graphical Methods of Statistical Inference*, Maruzen Co. Ltd.

Mathai, A.M. and Rathie, P.N., (1977), *Probability and Statistics*, S.G. Wasani for McMillan Co. of India Ltd.

Miller, R.L. and Kahn, S., (1962), *Statistical Analysis in Geological Sciences*, John Wiley & Sons.

Misra, R.P. (1989) *Fundamental of Cartography*, Concept Publishing Co., New Delhi.

Misra, R. P. (1989) *Research Methodology : A Hand Book*, Concept Publishing Co. New Delhi.

Monkhouse, F.J. and Wilkinson, H.R., (1964), *Maps and Diagrams—Their Compilation and Construction*, Methuen & Co. Ltd.

Mulholland, H. and Jones, C.R., (1975), *Fundamentals of Statistics*, English Language Book Society and Butterworth.

Murthy, M.N., (1967), *Sampling—Theory and Methods*, Statistical Pub. Society.

Maxwell, Albert, E., (1961), *Analysing Quantitative Data*, John Wiley & Sons, New York.

Moroney, M.J., (1956), *Facts from Figures*, Penguin Books, Baltimore.

Morrison, Donald, F., (1967), *Multivariate Statistical Methods*, McGraw Hill, New York.

Nie, N.H.; Bent, D.H. & Hull, C.H., (1970), *Statistical Package for the Social Sciences*, McGraw Hill, New York.

Ostle, B., (1966), *Statistics in Research*, Iowa State University Press.

Oppenheim, A.N., (1966), *Questionnaire Design and Attitude Measurement*, Basic Books, New York.

Ostle, Bernard and Mensing, R.W., (1975), *Statistics in Research*, The Iowa State University Press.

Payne, Stanley (1951), *The Art of Asking Questions*, Princeton University Press.

Rafiullah, S.M., (1965), A New Approach to Functional Classification of Towns, *The Geographer*, Vol. XII, p. 40.

Rao, P. and Miller, R.L., (1972), *Applied Econometrics*, Prentice Hall of India Pvt. Ltd.

Reynold, R.B., (1956), Statistical Methods in Geographical Research, *The Geographical Review*, p. 129.

Robinson, A.H. and Sale, R.D., *Elements of Cartography*, John Wiley and Sons Inc.

Ramachandran, P., (1971), Training in Research Methodology in *Social Sciences in India*, ICSSR, New Delhi.

Redman, L.W. and Mory, A.V.H., (1923), *Romance of Research*.

Sadhu, A.N. and Singh Amarjit, (1980), *Research Methodology in Social Sciences*, Himalyan Pub. House, Bombay.

Sharma, B.A.V. *et. al.*, (1983), *Research Methods in Social Sciences*, Sterling Pub., New Delhi.

Silk, J., (1979), *Statistical Concept in Geography*, George Allen and Unwin.

Simpson G. and Kafka, F. (1965), *Basic Statistics—Text Book for the First Course*, Oxford and IBH Publishing Co.

Sukhatme, P.V. and Sukhatme, B.V., (1970), *Sampling Theory With Applications*, Asia Pub. House, New Delhi.

Summer, G.N., (1978), *Mathematics for Physical Geographers*, Edward Arnold Pub. Ltd.

Theakstone, W.H. and Harison, C., (1970), *The Analysis of Geographical Data*, Heineman Edu. Books.

Unwin, D., (1981), *Introductory Spatial Analysis*, Methuen and Co.

Walker, H.M. and Lev. J., (1965), *Elementary Statistical Methods*, Oxford & IBH Pub. Co.

Weaver, J.C., (1964), Crop-Combination Regions in the Middlewest, *Geographical Review*, Vol. 44, p. 175.

Wilson, A.G. & Kirkby, M.J., (1975), *Mathematics for Geographers and Planners*, Clarendon Press.

Wood, W.F., (1955), Use of Stratified Random Samples in a Landuse Study, *Annals of the Association of American Geographers*, Vol. 45, p. 350.

Yeats, M.H., (1968), *An Introduction to Quantitative Analysis in Economic Geography*, McGraw Hill Book Co.

Index